English for Academic Research

C000242105

More information about this series at http://www.springer.com/series/13913

Adrian Wallwork

English for Academic Research: Writing Exercises

 Springer

Adrian Wallwork
Via Carducci 9
56127 Pisa, Italy
adrian.wallwork@gmail.com

ISBN 978-1-4614-4297-4   ISBN 978-1-4614-4298-1 (eBook)
DOI 10.1007/978-1-4614-4298-1
Springer New York Heidelberg Dordrecht London

Library of Congress Control Number: 2012948774

Printed on acid-free paper

Springer is part of Springer Science+Business Media (www.springer.com)

Preface

Aim of the book and coverage

The book is aimed at postgraduate students, PhD students and researchers whose first language is not English. It is assumed that you have already reached a sufficient level of English to write a research paper, thesis or dissertation.

The book covers all the writing skills that will help you to get a positive reaction from the reviewers of your manuscript, and thus improve your chances of publication. When reviewers say that the level of the English in a manuscript is 'poor', they are often referring not to grammar or vocabulary issues, but to readability problems (see the second section on page vii), such as poor structure, sentences being too long, redundancy, and ambiguity. All these problems, and many more, are dealt with in this book.

Structure of the book, self–study and classroom use

Sections 1–9 of the book practice particular writing skills. Section 10 brings all these skills together in exercises on writing specific sections of a manuscript – from the Abstract to the Acknowledgements. Around half of the exercises in Sections 1–9 can be done rapidly, without the aid of a teacher. They are thus suitable for self study. Other exercises require you to write extended pieces of text, which you will need to have corrected by your teacher or a native speaker of English. Each part begins with cross references to other books in the series (see the first section on page vii).

Instructions and keys to the exercises

Instructions to exercises are in *italics*. Examples of how to do the exercises are shaded in grey.

If there is no example given and you are not sure how to do the exercise, look at the first question in the exercise and then the answer to that question in the key.

The keys (solutions) to the exercise appear immediately below the exercise, but in a smaller font. The idea is that you don't have to flip to the back of the book to find the answers. This should speed up the process of doing the exercises. In a few cases, there is no key because there are unlimited ways of answering the exercise.

In any case, you should consider the keys as being suggested answers. There may be several possible answers. If in doubt, consult with your English teacher.

Word and phrases in [parentheses] indicate that these are alternative solutions to the ones outside parentheses, but they are probably less commonly used.

Word and phrases separated by a slash (e.g. *which/that*) indicate that both solutions are equally valid.

Language and 'facts' used in this book

A few of the texts may contain technical language that you may not be familiar with. However, it is not necessary to understand every word in each sentence in order to be able to do the exercise. But if you find that the technical language of one particular exercise is an obstacle to you being able to complete the exercise, then simply ignore that exercise and do the next. In fact, the book has been designed to give practice of the same writing skill in more than one exercise.

Most of the facts, statistics and authors' names contained in the exercises have been invented. Some are designed to be humorous. Academic writing can be quite heavy and you may find you are more motivated to do some exercises if there is an element of fun involved. You are thus encouraged to invent data and information. All the exercises reflect the typical style of academic works and many are based on real texts. So whether you are using true facts or inventing your own, the kind of language and constructions you use will be in the same academic style.

Cross referencing with other books in the series

This book is divided into ten parts. At the beginning of each part is a list of the writing skills practiced in the exercises. These skills are cross referenced to two other books in the series:

English for Research: Grammar, Usage and Style – designed to resolve your doubts about the grammar, usage and style of academic English.

English for Writing Research Papers – everything you need to know about how to write a paper that referees will recommend for publication.

This means that you can check how to use a particular writing skill before you start doing the related exercise. Grammar (e.g. the use of articles and tenses) and vocabulary are covered in:

English for Academic Research: Grammar Exercises
English for Academic Research: Vocabulary Exercises

Other books in the series that you might find useful are:

English for Academic Correspondence – tips for responding to editors and referees, networking at conferences, understanding fast-talking native English speakers, using Google Translate, and much more. No other book like this exists on the market.

English for Presentations at International Conferences – all the tricks for overcoming your fear of presenting in English at a conference.

English for Interacting on Campus – tips for: socializing with fellow students, addressing professors, participating in lectures, improving listening skills and pronunciation, surviving in a foreign country.

To find out how the manuals are cross-referenced with the exercise books go to: http://www.springer.com/series/13913

Focus on readability

Your main aim is to get your paper published. The people who determine whether your paper will be published are the editors of the journal and the referees who review your paper.

Readability. This is the key concern of referees. If a paper is not readable it cannot be published. If a paper contains a limited number of grammatical and lexical errors, it can still be published, because such errors rarely prevent the reader from understanding the paper.

Readability relates to the amount of effort the reader has to make in order to understand what you have written. Readability is affected by the following factors:

- sentence length (sentences longer than 30 words are generally hard to assimilate without having to be read twice)
- lack of structure (within a sentence, paragraph or section)
- redundancy (i.e. words, phrases, sentences, paragraphs and sections that add no value for the reader)
- ambiguity and lack of clarity (i.e. the reader is not sure how to interpret a phrase)

A low level of readability is associated with authors who are more interested in expressing themselves in an 'elegant' or 'academic' way, rather than on focusing on what the reader really wants/needs to read, and the best way to make this information immediately clear to the reader. English has increasingly become a reader-oriented language, in which authors feel a responsibility to help their readers, rather than impress them. This does not mean that English has become a simple language and that it has limited expressive power. It means that, when it is written well, it cuts out any unnecessary information, and presents all the useful information in a way that clearly shows the connections between ideas. Ideally, it does not leave gaps for the reader to fill in, nor does it adopt vague language and thus force the reader to make interpretations. Bear in mind, however, that there are still many native English writers whose aim seems to be to obscure rather than enlighten!

Think about what you like reading on the web. You probably appreciate:

- ease in finding the information you want
- short sentences and paragraphs containing only relevant information
- white space, no dense blocks of text
- no distractors (e.g. pop ups, animations, links in every other sentence)

When you write your paper, bear the above in mind. Think about what you like reading, then try to write in a style that will make reading your paper a pleasurable experience for your audience. Make it easy for readers to find what they want and to absorb it. Don't create distractors: so no redundant words and phrases, misspellings, pointless or difficult tables and figures. And don't make your readers wait for key information or force them to read something twice before they can understand it.

A note for teachers

This book of exercises is designed to be used in conjunction with *English for Writing Research Papers*, which is part of the same series of books.

I have tried to cover what I consider to be the most important aspects of writing, particularly the ones that are likely to cause a paper to be rejected. Exercises on grammar and vocabulary can be found in the other volumes of this series.

Many of the exercises, particularly those in Chapters 1-5, can be set as homework as they are quick to do and contain a key. The key is on the same page as the exercise. Simply tell the students to cover the key while they are doing the exercise.

Also the extended exercises (e.g. those in Chapter 10) can be done at home.

I suggest that you use classtime to:

- explain the theory (you can prepare by yourself using the relevant sections from *English for Writing Research Papers*)
- go over the exercises

For full details on how to exploit all the books in the English for Academic series, see:

English for Academic Research: A Guide for Teachers

Contents

1   Punctuation and spelling.. 1

   1.1   commas: reducing number of.. 2

   1.2   commas: adding 3

   1.3   semicolons: replacing... 4

   1.4   brackets: removing 5

   1.5   hyphens: adding 6

   1.6   hyphens: deciding where needed................................... 7

   1.7   initial capitalization: in titles 8

   1.8   initial capitalization: in main text 9

   1.9   various punctuation issues: 1 10

  1.10   various punctuation issues: 2 11

  1.11   spelling 12

2   Word order 13

   2.1   choosing the best subject to put at the beginning

         of the phrase 14

   2.2   putting the key words first... 16

   2.3   avoiding beginning the sentence with *it is*: 1 17

   2.4   avoiding beginning the sentence with *it is*: 2 18

   2.5   choosing the best word order to help the reader: 1........... 19

   2.6   choosing the best word order to help the reader: 2........... 22

   2.7   choosing the best word order to help the reader: 3........... 24

   2.8   shifting the parts of the phrase to achieve

         optimal order: 1 25

   2.9   shifting the parts of the phrase to achieve

         optimal order: 2 26

  2.10   shifting the parts of the phrase to achieve

         optimal order: 3 28

  2.11   shifting the parts of the phrase to achieve

         optimal order: 4 29

2.12   reducing the number of commas and parts
       of the sentence.. 30
2.13   putting sentences into the correct order...................... 31
2.14   typical mistakes... 32

3   Writing short sentences and paragraphs 33
3.1   dividing up long sentences: 1 34
3.2   dividing up long sentences: 2 36
3.3   dividing up long paragraphs 1 38
3.4   dividing up long paragraphs 2 40
3.5   dividing up long paragraphs 3 42
3.6   dividing up long paragraphs 4 43
3.7   putting paragraphs into their most logical order 45
3.8   writing short sentences: 1.. 46
3.9   writing short sentences: 2.. 46
3.10   writing short sentences: 3.. 46

4   Link words: connecting phrases and sentences together 47
4.1   linking sentences and paragraphs............................. 48
4.2   deleting unnecessary link words 49
4.3   deciding when link words are necessary.................... 50
4.4   choosing best link word.. 51
4.5   reducing the length of link words / phrases 52
4.6   shifting the position of link words expressing
       consequences 53
4.7   using link words to give additional neutral information 54
4.8   using link words to give additional positive information 55
4.9   using link words to give additional negative information..... 56
4.10   making contrasts 57
4.11   making evaluations.. 58
4.12   connecting sentences by repetition of key word
       or a derivation of the key word 59
4.13   describing processes 60
4.14   describing causes 61
4.15   describing effects and consequences 61
4.16   making contrasts, concessions, qualifications,
       reservations, rejections 62
4.17   outlining solutions to problems.................................. 63
4.18   outlining a time sequence.. 65
4.19   explaining figures and tables: making comparisons 66
4.20   making evaluations and drawing conclusions: 1.................. 67
4.21   making evaluations and drawing conclusions: 2................. 67

5   Being concise and removing redundancy................... 69
5.1   removing individual redundant words.......................... 70
5.2   removing several redundant words: 1.......................... 71
5.3   removing several redundant words: 2.......................... 73

5.4   reducing the word count: titles.. 75
5.5   replacing several words with one preposition
       or adverb 76
5.6   replacing several words with one adverb........................ 77
5.7   replacing several words with one word........................... 78
5.8   replacing a *verb + noun* construction with
       a single verb: 1 79
5.9   identifying verb and noun clauses 80
5.10   replacing a *verb + noun* construction with
       a single verb: 2 82
5.11   replacing a noun phrase with a verb or *can*: 1.................... 83
5.12   replacing a noun phrase with a verb or *can*: 2.................... 84
5.13   replacing nouns with verbs in titles of papers................... 85
5.14   identifying whether link words could be deleted 86
5.15   deleting unnecessary link words: 1 87
5.16   deleting unnecessary link words: 2 88
5.17   deleting unnecessary link words: 3 89
5.18   unnecessary use of *we* and *one:* 1 90
5.19   unnecessary use of *we* and *one:* 2 91
5.20   avoiding redundancy in introductory phrases.................. 92
5.21   avoiding redundancy in references to figures,
       tables etc. 92
5.22   rewriting unnecessarily long sentences: 1........................ 93
5.23   rewriting unnecessarily long sentences: 2........................ 94
5.24   rewriting unnecessarily long sentences: 3........................ 96
5.25   reducing length of an abstract...................................... 97
5.26   reducing length of an introduction 98
5.27   reducing the length of the outline of the structure 99
5.28   reducing the length of the review of the literature: 1.......... 100
5.29   reducing the length of the review of the literature: 2.......... 101
5.30   reducing the length of the materials and methods 102
5.31   reducing the length of the conclusions section.................. 102
5.32   reducing the length of the acknowledgements 103

6   **Ambiguity and political correctness** 105
6.1   repetition of words to aid reader's understanding: 1.......... 106
6.2   repetition of words to aid reader's understanding: 2.......... 107
6.3   avoiding ambiguity due to use of *-ing* form: 1 108
6.4   avoiding ambiguity due to use of *-ing* form: 2.................... 109
6.5   disambiguating sentences: 1... 110
6.6   disambiguating sentences: 2... 112
6.7   pronouns and political correctness................................. 113
6.8   non-use of masculine terms for generic situations: 1 114
6.9   non-use of masculine terms for generic situations: 2 115
6.10   non-use of masculine terms for generic situations: 3 115

7   Paraphrasing and avoiding plagiarism 117

  7.1   deciding what is acceptable to cut and paste 118
  7.2   quoting statistics 119
  7.3   paraphrasing by changing the parts of speech 120
  7.4   paraphrasing by changing nouns into verbs 121
  7.5   paraphrasing by changing the parts of speech
        and word order: 1 123
  7.6   paraphrasing by changing the parts of speech
        and word order: 2 124
  7.7   finding synonyms: verbs 1 125
  7.8   finding synonyms: verbs 2 126
  7.9   finding synonyms: verbs 3 127
  7.10 finding synonyms: nouns 1 128
  7.11 finding synonyms: nouns 2 129
  7.12 finding synonyms: adjectives 130
  7.13 finding synonyms: adverbs and prepositions 1 131
  7.14 finding synonyms: adverbs and prepositions 2 132
  7.15 paraphrasing by changing word order 133
  7.16 replacing *we* with the passive form 134
  7.17 making a summary: 1 136
  7.18 making a summary: 2 137
  7.19 making a summary: 3 138
  7.20 making a summary: 4 138

8   Defining, comparing, evaluating and highlighting 139

  8.1   writing definitions 1 140
  8.2   writing definitions 2 141
  8.3   writing definitions 3 141
  8.4   making generalizations 142
  8.5   confirming other authors' evidence 143
  8.6   stating how a finding is important 144
  8.7   highlighting why your method, findings, results etc.
        are important 146
  8.8   highlighting your findings 147
  8.9   comparing the literature 148
  8.10 comparing contrasting views 151
  8.11 comparing your methodology with other authors'
        methodologies 152
  8.12 comparing data in a table 153
  8.13 questioning current thinking 154
  8.14 evaluating solutions 155

**9   Anticipating possible objections, indicating level of certainty,
      discussing limitations, hedging, future work** 157
   9.1   anticipating objections and alternative views.................. 158
   9.2   indicating level of certainty 1 159
   9.3   indicating level of certainty 2 160
   9.4   reducing level of certainty.. 161
   9.5   discussing the limitations of the current state
        of the art 162
   9.6   qualifying what you say 162
   9.7   dealing with limitations in your own results: 1 163
   9.8   dealing with limitations in your own results: 2 164
   9.9   dealing with limitations in your own results: 3 165
   9.10   toning down the strength of an affirmation: 1 166
   9.11   toning down the strength of an affirmation: 2 167
   9.12   toning down the strength of an affirmation: 3 168
   9.13   toning down the strength of an affirmation: 4 169
   9.14   direct versus hedged statements 1.................................. 170
   9.15   direct versus hedged statements 2.................................. 171
   9.16   discussing possible applications and future work............ 173

10   Writing each section of a paper 175
   10.1   abstracts... 176
   10.2   introductions 177
   10.3   creating variety when outlining the structure
        of the paper 177
   10.4   outlining the structure of the paper................................. 178
   10.5   survey of the literature.. 178
   10.6   methodology / experimental 179
   10.7   results... 180
   10.8   discussion: 1 180
   10.9   discussion: 2 181
   10.10 differentiating between the abstract
        and the conclusions: 1 182
   10.11 differentiating between the abstract
        and the conclusions: 2 183
   10.12 conclusions: 1 184
   10.13 conclusions: 2 184
   10.14 acknowledgements: 1 185
   10.15 acknowledgements: 2 185

Acknowledgements.. 187

About the Author 187

**Editing Service for non-native researchers / Mentorship
    for EAP and EFL teachers** 187

Index 189

Section 1: Punctuation and spelling

Abstract

TOPIC	ENGLISH FOR RESEARCH USAGE, STYLE, AND GRAMMAR	ENGLISH FOR WRITING RESEARCH PAPERS
apostrophes	25.1	
colons	25.2	
commas	25.3, 25.4	3.14
hyphens	25.6, 25.7	
parentheses / brackets	25.8	3.17
periods (full stops)	25.9	
semicolons	25.11	3.15, 3.16
titles		11.6, 11.7

A. Wallwork, *English for Academic Research: Writing Exercises*,
DOI 10.1007/978-1-4614-4298-1_1, © Springer Science+Business Media New York 2013

1.1 commas: reducing number of

Reduce the number of commas in the following sentences by changing the phrase round.

The specimens, each of which was cruciform, weighed 90–100 g..

= The specimens were cruciform and weighed 90–100 g.

1. This device, as is well known, will separate X from Y.

2. Let us strengthen, by means of the circuit of Fig. 3b, the example given in the previous subsection.

3. This book, which is aimed at non native researchers contains a series of exercises practising writing skills.

4. This paper, which is an amplified version of a paper presented at the Third Conference on Writing, is divided into five main parts.

5. The results of the survey, once they have been processed, will be used to make a full assessment of the advantages of such an approach.

6. Let us take into consideration, using the data given in Table 1, the most important parameters.

1. As is well known, this device will separate X from Y.

2. Using the circuit of Fig. 3b, let us strengthen the example given in the previous subsection.

3. This book is aimed at non native researchers. It contains a series of exercises practising writing skills.

4. This paper is an amplified version of a paper presented at the Third Conference on Writing, and is divided into five main parts.

5. Once the results of the survey have been processed, they will be used to make a full assessment of the advantages of such an approach.

6. Using the data given in Table 1 the most important parameters can be considered.

1.2   commas: adding

Insert commas where needed. Do not change the order of the words.

X not Y was the most useful.

= X, not Y, was the most useful.

1. Artists have always experimented with a variety of organic natural materials for use as paint binders and varnishes and as ingredients for mordants.

2. First he spoke about X and then about Y.

3. For breakfast I have yoghurt corn flakes and bacon and egg.

4. In direct inlet mass spectrometry solid or liquid samples are introduced into a small glass cup.

5. It costs $2200000. ← It's a number

6. More and more Americans wait until the deadline to pay their bills.

7. The menu offered the usual choices of turkey lamb and chicken.

8. The paper is in three parts. Firstly we look into X. Secondly we discuss Y. Thirdly we investigate Z.

The commas are marked with a hash (#).

1. Artists have always experimented with a variety of organic natural materials for use as paint binders and varnishes # and as ingredients for mordants.

2. First he spoke about X # and then about Y.

3. For breakfast I have yoghurt # corn flakes # and bacon and egg.

4. In direct inlet mass spectrometry # solid or liquid samples are introduced into a small glass cup.

5. It costs $2 # 200 # 000.

6. More and more # Americans wait until the deadline to pay their bills. (*more and more* has been interpreted as meaning *increasingly*).

7. The menu offered the usual choices of turkey # lamb and chicken.

8. The paper is in three parts. Firstly # we look into X. Secondly # we discuss Y. Thirdly # we investigate Z.

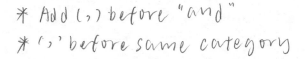

* Add (,) before "and"

* ',' before same category

4

1.3 semicolons: replacing

*Where possible and appropriate, remove the semicolons and replace them
with full stops or commas.* ~~towns of a place~~

One relatively easy method to collect information is asking for it directly
from the individual under a form of questionnaire. I used questionnaires
from bilingual subjects located in Florence, **(1) Tuscany;** Milan, → (,)
← **(2) Lombardy;** and Rome, Lazio. The participants were asked, for (,)
example, when they acquired their second **(3) language;** if they use both
← languages **(4) regularly;** and how they self-rate their level of proficiency in (,)
reading, listening, writing and speaking. Issues may arise from a different importance given to factors affecting language **(5) acquisition;** for → (.)
example, can the length of residence in a foreign country be considered
an index of proficiency? Some 27 attempts have been made to develop a
reliable and valid questionnaire, which could predict the relationships with
objective measures (e.g., Marian, Blumenfeld & Kaushanskaya, **(6) 2007;**
Tokowicz, Michael & Kroll, 2004). Although these questionnaires were
all different, they showed a consistent degree of overlapping items, for
example, age of L2 (i.e. second language) first **(7) exposure;** years of L2 → (,)
← instruction **(8) received;** and language spoken at home. Li et al. (2006) (,)
identified these recurring items, which were consolidated into a single
← **(9) source;** however, despite the authors' intention to add new functions (.)
to the interface they did not develop it further.

1 and 2: semicolons are required here as otherwise the reader would not be able to
distinguish the towns (Florence, Milan, Rome) from their regions (Tuscany etc.) – it would
seem that all the words were related to the same type of entity.

3 and 4: semicolons could be replaced by a comma as there is no possible confusion here.

5: a full stop could be used here.

6: the semicolons are useful to divide up the list of authors into separate groups.

7 and 8: semicolons are not strictly necessary here, but they help the reader differentiate
between the commas (which in this case are used to separate this long sentence into
shorter clauses) and the semicolons which divide up the items in the list.

9: a full stop should be used here.

1.4   brackets: removing

Where possible, remove the brackets and rewrite the sentences accordingly. In some cases, the information in brackets is redundant.

1. If the method is used correctly (i.e. each step is followed in sequence) then the results are generally in line with the best results obtained by other authors in the literature (e.g. Smith [2011], Yang [2012] and Singh [2013]).

2. For certain countries (e.g. Peru, Chile and Honduras) these distinctions do no apply.

3. The software performs all the checks (price, quantity, quality).

4. In practice this allows the users to shift the queries to one direction or the other, depending on a (positive or negative) skewing factor.

5. We decided to use this procedure (as defined in ISO 12 / 998) as it is generally more efficient.

1. If each step in the method is followed in sequence then the results are generally in line with the best results obtained by other authors in the literature, for example Smith [2011], Yang [2012] and Singh [2013].

2. For certain countries (e.g. Peru, Chile and Honduras) these distinctions do no apply.

3. The software checks the price, quantity, and quality.

4. In practice this allows the users to shift the queries to one direction or the other, depending on a skewing factor that can either be positive or negative.

5. We decided to use this procedure (ISO 12 / 998) as it is generally more efficient.

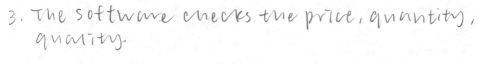

3. The software checks the price, quantity, quantity.

5. We decided to use

1.5 hyphens: adding

Insert hyphens where needed.

She has a full time job. = She has a full-time job.

1. An Italo American project.
2. A second order problem.
3. A 50 year old man.
4. A 10 year period.
5. The use of a compiler controlled network.
6. Via point to point routing.
7. On the fly compilation.
8. We need to look at the decision making process.
9. This is not a heart related illness.
10. There is an ever growing need for such devices.
11. These are all real life situations.
12. This entails using a market based mechanism.
13. Our profit maximizing models solve this problem very neatly.
14. It is a robot like device.
15. There is no mention of any time dependent factors.

1. An Italo-American project.
2. A second-order problem.
3. A 50-year-old man.
4. A 10-year period.
5. The use of a compiler-controlled network.
6. Via point-to-point routing.
7. On-the-fly compilation.
8. We need to look at the decision-making process.
9. This is not a heart-related illness.
10. There is an ever-growing need for such devices.
11. These are all real-life situations.
12. This entails using a market-based mechanism.
13. Our profit-maximizing models solve this problem very neatly.
14. It is a robot-like device.
15. There is no mention of any time-dependent factors.

1.6 hyphens: deciding where needed

Decide in which sentence, a or b, hyphens are required between the words in bold. In two cases hyphens are required in both a and b, but in different positions and consequently with different meanings.

1 (a) These spread from **cell to cell**.

1 (b) **Cell to cell** communication is frequent.

2 (a) This behavior is **human like**.

2 (b) This is a **human like** behavior.

3 (a) This is a **well known** problem.

3 (b) This problem is **well known**.

4 (a) We will review the **state of the art** in the literature.

4 (b) This is a **state of the art** piece of equipment.

5 (a) This is used to **clean up** the sample.

5 (b) Do this after the **clean up** procedure.

6 (a) There is a **one to one** correspondence.

6 (b) These should be done **one by one**.

7 (a) A traditional **single cluster** assignment.

7 (b) There is just a **single cluster**.

8 (a) These students are in their **third year**.

8 (b) These are **third year** students.

9 (a) This is a **little used car** – it is very compact.

9 (b) This is a **little used car** – it has only done 2000 km.

10 (a) We approached several **foreign car dealers** (e.g. Ferrari, Honda, Kia) who told us …

10 (b) We approached several **foreign car dealers** (i.e. not from the USA) who told us …

1	(b)	6	(a)
2	(b)	7	(a)
3	(a)	8	(b)
4	(b)	9	(a) Used-car (b) Little-used
5	(b)	10	(a) Foreign-car (b) Car-dealers

1.7 initial capitalization: in titles

Use initial capitalization on the following titles.

Consequences of erudite vernacular utilized irrespective of necessity: the problems of using long words needlessly.

= Consequences of Erudite Vernacular Utilized Irrespective of Necessity: the Problems of Using Long Words Needlessly.

1. A guide to writing research papers for non-native speakers of English.

2. The role of English in the twenty-first century.

3. The history of teaching English as a foreign language.

4. An innovative system for the automatic translation of research papers.

1. A Guide to Writing Research Papers for Non-Native Speakers of English.

2. The Role of English in the Twenty-First Century.

3. The History of Teaching English as a Foreign Language.

4. An Innovative System for the Automatic Translation of Research Papers.

1.8   initial capitalization: in main text

Underline any words that should begin with an initial capital letter.

1. The values are shown in table 1. This table also shows the daily doses from monday to friday.

2. The authors gratefully acknowledge support from the university of manchester. Thanks are also due to dr susan james for revising the english of the manuscript.

3. In order to maximize background conductivity, a dionex anion micro membrane suppressor (dionex, sunnyvale, usa) was employed.

4. This paper introduces logibase, a system that integrates a spreadsheet, a relational data base, and logic programming paradigms by exploiting boolean values.

1. The values are shown in *Table* 1. This table also shows the daily doses from *Monday* to *Friday*.

2. The authors gratefully acknowledge support from the *University* of *Manchester*. Thanks are also due to *Dr Susan James* for revising the *English* of the manuscript.

3. In order to maximize background conductivity a *Dionex* anion micro membrane suppressor (*Dionex, Sunnyvale, USA*) was employed.

4. This paper introduces *Logibase*, a system that integrates a spreadsheet, a relational data base, and logic programming paradigms by exploiting *Boolean* values.

1.9 various punctuation issues: 1

Insert punctuation (including capitalization) into the following text which is part of an Introduction of an informal paper. Note that some sentences are very short.

The order in which we say or write something generally reflects the importance we want to give to each individual item in english we tend to put the subject first because by doing this the interlocutor immediately knows what the main topic is going to be we then need to insert the verb and then the object which is generally of secondary importance this may seem obvious it isnt in many languages the subject or a part of the verb may appear at the end of the sentence this fact would seem to indicate that we dont all have the same thought patterns and that for other nationalities it may not be important to immediately know the argument of the sentence.

The result is that we as listeners or readers have certain expectations as to the order in which the words are going to appear if this order is not respected we may be thrown off the trail in much the same way foreign students when speaking tend to stress inappropriate words in a sentence highlighting words that for an english speaker would normally have no relevance the reason for this is that both english word order and english stress are strictly related to meaning in other languages this is not always the case the problem is of course that students tend to transfer their native word and stress into english.

NB there are other possible ways of punctuating these two paragraphs.

The order in which we say or write something generally reflects the importance we want to give to each individual item. In English we tend to put the subject first because by doing this the interlocutor immediately knows what the main topic is going to be. We then need to insert the verb; and then the object, which is generally of secondary importance. This may seem obvious. It isn't. In many languages the subject or a part of the verb may appear at the end of the sentence. This fact would seem to indicate that we don't all have the same thought patterns, and that for other nationalities it may not be important to immediately know the argument of the sentence.

The result is that we, as listeners or readers, have certain expectations as to the order in which the words are going to appear. If this order is not respected, we may be thrown off the trail. In much the same way foreign students, when speaking, tend to stress inappropriate words in a sentence, highlighting words that for an English speaker would normally have no relevance. The reason for this is that both English word order and English stress are strictly related to meaning. In other languages this is not always the case. The problem is of course that students tend to transfer their native word and stress into English.

1.10   various punctuation issues: 2

Insert punctuation (including parentheses, hyphens and capitalization) into this Methods section from a medical paper. The 'clients' referred to are homeless people.

The homeless population involved in the study include those in temporary or insecure housing in a hostel staying with friends or relatives out of necessity or sleeping rough. clients were screened at thirty five sites which make up the main hostels, nightshelters and day centres for homeless people and refugees in the south london boroughs of lambeth, lewisham and southwark many of these agencies target homeless people who do not normally access other services particularly health care the screening was advertised in advance at each site as a free service available to all with incentives free meals provided regular screening sessions were arranged at each site over a number of months the frequency of sessions depended on the size of the centre and the daily turnover of clients the overall uptake of the screening at each site was estimated by calculating the average number of volunteers for the screening at each centre as a percentage of the average daily capacity of each drop in centre or hostel.

The homeless population involved in the study include those in temporary or insecure housing, in a hostel, staying with friends or relatives out of necessity, or sleeping rough. Clients were screened at thirty-five sites, which make up the main hostels, nightshelters and day centres for homeless people and refugees in the south London boroughs of Lambeth, Lewisham and Southwark. Many of these agencies target street homeless people who do not normally access other services (particularly health care). The screening was advertised in advance at each site as a free service, available to all, with incentives (free meals) provided. Regular screening sessions were arranged at each site over a number of months. The frequency of sessions depended on the size of the centre and the daily turnover of clients. The overall uptake of the screening at each site was estimated by calculating the average number of volunteers for the screening at each centre as a percentage of the average daily capacity of each drop-in centre or hostel.

1.11 spelling

Choose the correct spelling. Note that in some cases, two spellings are possible (one for Great Britain and one for the USA).

1. Section 1 outlines our **preferred / prefered** mode of operation. Section 2 is **targeted / targetted** at inexpert users. In Section 5, **were / where** we suggest some future lines of research, **some / same** conclusions are **drown / drawn**.

2. The samples were **weighed / weighted**. The **weighed / weighted** values were processed.

3. We believe that in order to **fullfil / fulfill / fulfil** these objectives we need extra funding, **though / tough** another approach might be to **find / found** sponsors.

4. The **price / prize** of this approach is that it is very much **dependant / dependent** on how the file is **attacked / attached** to the email.

5. At the **beginning / begining** I was looking for **accomodation / accommodation** in the **center / centre**, then I **found / founded** a flat in the suburbs.

6. It would be **usefull / useful** to do this as a **separate / seperate** activity rather than in **paralel / parallel**.

7. I need to improve my **pronunciation / pronounciation** because it makes me **embarrassed / embarassed** when I make a mistake.

8. There are also some **constrains / constraints** that need to be dealt with **form / from** the point of view of the performance of the **aluminium / aluminum** casing.

9. The **aging / ageing** process includes hair going **grey / gray**, less chances of finding **labour / labor**, becoming more **sceptical / skeptical**, and possibly being more prone to **diarrhoea / diarrhea**, but increased time and money for **traveling / travelling** and going to the **theatre / theater**.

10. **Acknowledgements / acknowledgments**: The authors would like to **tank / thank** the following people …

1. preferred, targetted, where, some, drawn

2. weighed, weighted

3. fulfill (US) / fulfil (GB), though, find

4. price, dependent, attacked

5. beginning, accommodation, center (US) / centre (GB), found

6. useful, separate, parallel

7. pronunciation, embarassed

8. constraints, from, aluminium (GB) / aluminum (US)

9. aging / ageing, grey / gray, labour (GB) / labor (US), sceptical (GB) / skeptical (US), diarrhoea (GB) / diarrhea (US), travelling (GB) / traveling (US), theatre (GB) / theater (US)

10. acknowledgements (GB) / acknowledgments (US), thank

Section 2: Word order

TOPIC	ENGLISH FOR RESEARCH USAGE, STYLE, AND GRAMMAR	ENGLISH FOR WRITING RESEARCH PAPERS
adjectives[a]	18.1–18.3, 19.2, 19.4	2.13, 2.14
adverb[a]	Section 17	2.12
avoiding putting it at the beginning of the sentence		2.5, 2.6, 5.11
basic word order		2.1
choosing the best subject	16.2	2.3, 2.4
direct and indirect objects[a]	16.8	2.9, 2.10
inversion of subject and verb	16.5–16.7	
parenthetical information	15.3	2.9
past participles	18.4	
subject	16.1–16.5	2.7
verb		2.8

[a]these are practised in more detail the companion volume: *English for Academic Research: Grammar Exercises*

A. Wallwork, *English for Academic Research: Writing Exercises*,
DOI 10.1007/978-1-4614-4298-1_2, © Springer Science+Business Media New York 2013

2.1 choosing the best subject to put at the beginning of the phrase

Choose the best sentence (a or b). The parts in bold are designed to show you the main differences in the word order in order to help you choose the best option.

(1a) **The following are** some examples of rare species:

(1b) Examples of rare species **are the following**:

(2a) Among the factors which influence longevity of seeds, of particular importance are **temperature and moisture content**.

(2b) **Temperature and moisture content** are particularly important factors influencing the longevity of seeds.

(3a) Sometimes 802.16 systems are referred to as Worldwide Interoperability for Microwave Access (WiMAX) systems **in the trade press**.

(3b) **In the trade press**, 802.16 systems are sometimes referred to as Worldwide Interoperability for Microwave Access (WiMAX) systems.

(4a) However, **this operation is only defined for some nouns**, which are called countable nouns.

(4b) However, **only for some nouns this operation is defined**, these nouns are called countable nouns.

(5a) **To do this exercise**, you do not need to be able to understand the meaning of the technical words.

(5b) You do not need to be able to understand the meaning of the technical words **in order to do this exercise**.

(6a) A gradual decline in germinability and in the subsequent vigor of the resultant seedling, a higher sensitivity to stresses upon germination and eventually loss of the ability to germinate **are generally recorded**.

(6b) **There is generally** a gradual decline in germinability and in the subsequent vigor of the resultant seedling, followed by a higher sensitivity to stress upon germination, and eventually a loss of the ability to germinate.

(7a) This leads to the expression **in the plasma membrane** of AGEs derived from misfolded proteins, which are known to transmit to surrounding cells (Fig. 2).

(7b) This leads to the expression of AGEs derived from misfolded proteins **in the plasma membrane**, which are known to transmit to surrounding cells (Fig. 2).

(1) a (this reflects the normal word order in English which is to put the subject of the verb at the beginning of the sentence)

(2) b (as in 1a)

(3) b (in the trade press is crucial information which should go first in the sentence)

(4) a (the construction in 4b is not correct English)

(5) a (for the same reasons as in 3)

(6) b (the verb in 7a is located at the end of the sentence, this means that the reader has to wait a long time before getting the key information contained in the verb)

(7) a (although 8a does not reflect the usual English construction of putting the direct object before the indirect object, it avoids the ambiguity of 8b where it seems that the proteins are in the plasma)

2.2   putting the key words first

Rewrite the sentences so that they begin with a subject. There are several possible ways to do this as highlighted in the example. Just choose one way. There is no key to this exercise as there are many possible solutions.

Concerning the role of education, it is given great emphasis in their society.
= The role of education is given great emphasis in their society.
= Their society gives great emphasis to the role of education.
= Great emphasis is given to the role of education in their society.

1. As far as religion is concerned, it plays a fundamental role.

2. With regard to politics, the vast majority of politicians are men.

3. In relation to performance, this increased in direct relation to the number of training sessions.

4. Concerning the side effects of the treatment, only one serious effect is currently known about.

5. Regarding the best way to learn a language, several theories have recently been developed.

2.3   avoiding beginning the sentence with *it is*: 1

Complete the second sentence so that it means the same as the first.

1. It is possible to use several strategies to achieve these goals.
   Several strategies ...

2. It is possible with this model to give the actual flow rate.
   This model gives ...

3. It is certain / sure that the new laws will benefit nuclear research.
   The new laws will

4. It is mandatory to use X.
   X is ...

5. It is possible to demonstrate [Kim, 2014] that ...
   Kim [2014] ...

6. It is anticipated there will be a rise in stock prices.
   A rise in stock prices ...

7. It is regretted that no funds will be available for the next academic year.
   Unfortunately, ...

1. Several strategies **can** be used to achieve these goals.

2. This model **gives** the actual flow rate.

3. The new laws will **certainly / surely** benefit nuclear research.

4. X is mandatory.

5. Kim demonstrated that ...

6. A rise in stock prices is anticipated.

7. Unfortunately, no funds will be ...

2.4 avoiding beginning the sentence with *it is*: 2

Rewrite these sentences so that they do not begin with *it is*.

1. It is easy to carry out these tests.

2. It is regretted that your manuscript does not fit the scope of the conference.

3. It is possible that salaries will increase.

4. It is important to clean the samples.

5. It is necessary to define the stresses with respect to the original configuration.

6. It is highly probable that all future implantations will be required to adhere to new safety rules.

7. It would be advisable to calculate the coefficients beforehand.

8. It is reasonable to think that at least one value will equal X.

9. It is possible to use the code for other purposes as well.

1. Carrying out these tests is easy. / These tests are easy to carry out.

2. Unfortunately, your manuscript does not fit the scope of the conference.

3. Salaries may increase.

4. The samples must be cleaned.

5. The stresses should be defined with respect to the original configuration.

6. All future implantations are very likely to be required to adhere to new safety rules.

7. The coefficients should be calculated beforehand.

8. At least one value will probably equal X.

9. The code can be used for other purposes as well.

2.5 choosing the best word order to help the reader: 1

Choose the sentence (a, b or c) that best enables the reader to quickly assimilate the information contained in the sentence.

Examples

(a) This makes it possible to read with sufficient precision the sensor.

(b) This makes it possible to read the sensor with sufficient precision. yes.

(a) Our aim was to assess the contribution both in the past and the present of anthropogenic activities to global environmental pollution.

(b) Our aim was to assess the contribution of anthropogenic activities to global environmental pollution both in the past and the present. yes.

Key: In both cases a) is incorrect because it interrupts the normal word order of English: subject + verb + object

(1a) Do you have any openings in your laboratory for PhD students?

(1b) Do you have any openings for PhD students in your laboratory?

(1c) For PhD students do you have any openings in your laboratory?

(2a) We are planning at my department a series of workshops on XYZ in November this year.

(2b) At my department we are planning in November this year a series of workshops on XYZ.

(2c) At my department we are planning a series of workshops on XYZ in November this year.

(3a) I would like to request a delay in submission of manuscript #: 08SFL-00975 until 21 October.

(3b) Until 21 October I would like to request a delay in submission of manuscript #: 08SFL-00975.

(3c) I would like to request a delay until 21 October in submission of manuscript #: 08SFL-00975.

(4a) Please find attached a copy of the paper for your convenience.

(4b) For your convenience, please find attached a copy of the paper.

(4c) Please, for your convenience find attached a copy of the paper.

(5a) I inadvertently submitted my manuscript #08CV-0069 for the SAE Magnets Congress, as an "Oral only Presentation" instead of a "Written and Oral Presentation".

(5b) I inadvertently submitted for the SAE Magnets Congress my manuscript #08CV-0069, as an "Oral only Presentation" instead of a "Written and Oral Presentation".

(5c) I inadvertently submitted as an "Oral only Presentation" instead of a "Written and Oral Presentation" my manuscript #08CV-0069 for the SAE Magnets Congress.

(6a) Please could you let me know how I can change the status of my paper.

(6b) Please you could let me know how I can change the status of my paper.

(6c) Please could you let me know how can I change the status of my paper.

(7a) Given that our deadline is the first week of next month, I would be grateful to receive your revisions by the end of this month.

(7b) By the end of this month, given that our deadline is the first week of next month, I would be grateful to receive your revisions.

(7c) Given that our deadline is the first week of next month, I would be grateful to receive by the end of this month your revisions.

(8a) I have raised this problem twice before in fact as you can see from the attached emails below,

(8b) From the attached emails below, as you can see I have in fact raised this problem twice before.

(8c) As you can see from the attached emails below, I have in fact raised this problem twice before.

(9a) To speed the process up very much anything you could do would be appreciated.

(9b) Anything could you do to speed the process up would be appreciated very much.

(9c) Anything you could do to speed the process up would be very much appreciated.

(10a) I resubmitted my manuscript (ID 09–00236.R1), revised according to the Editor's and Referees' comments, on April 3 of this year.

(10b) Revised according to the Editor's and Referees' comments, on April 3 of this year I resubmitted my manuscript (ID 09–00236.R1).

(10c) On April 3 of this year I resubmitted my manuscript (ID 09–00236. R1), revised according to the Editor's and Referees' comments.

(1) b (a)

(2) c

(3) a

(4) a

(5) a

(6) a

(7) a

(8) b

(9) b

(10) c

2.6   choosing the best word order to help the reader: 2

Choose the sentence (a or b) that best enables the reader to quickly assimilate the information contained in the sentence.

(1a) The geothermal fields in Iceland represent a significant test site for assessing the robustness of such methods.

(1b) A significant test site for assessing the robustness of such methods is represented by the geothermal fields in Iceland.

(2a) A detailed analysis on samples was carried out in order to understand whether this anomaly was due to the extraction process and the resulting alterations and / or by the presence of fractures.

(2b) In order to understand whether this anomaly was due to the extraction process and the resulting alterations and / or by the presence of fractures a detailed analysis on samples was carried out.

(3a) The findings highlighted in patients with severe disabilities a lack of this kind of motor function.

(3b) The findings highlighted a lack of this kind of motor function in patients with severe disabilities.

(4a) The results of the experiments show a good quality of the prediction when high precision is required.

(4b) The results of the experiments show that the prediction is of a good quality when high precision is required.

(5a) Finally, the results gained during the last competition, in which the University of Seoul participated for the first time, confirm the reliability of the system.

(5b) Finally, the reliability of the system was confirmed by the results gained during the last competition, in which the University of Seoul participated for the first time.

(6a) The increase in power makes it possible to download the data with sufficient speed.

(6b) The increase in power makes it possible to download with sufficient speed the data.

(7a) The following equation describes the circuit:

(7b) The equation that describes the circuit is the following:

(8a) The novelty and possibilities, such as its use for making long-term analyses, of the approach are based on …

(8b) The novelty and possibilities of the approach, such as its use for making long-term analyses, are based on …

(9a) These factors since 2012 have been considered of primary importance.

(9b) Since 2012 these factors have been considered of primary importance.

(10a) This will avoid discharging around eight million tons of debris into the atmosphere in 2020.

(10b) This will avoid discharging into the atmosphere in 2020 around eight million tons of debris.

(1) a

(2) a

(3) b

(4) b

(5) b

(6) a

(7) a

(8) b

(9) b

(10) a

2.7   choosing the best word order to help the reader: 3

Where necessary, rearrange the elements in these sentences into the clearest order for the reader. Insert commas where necessary.

1. Among the factors which influence X / particularly important / are P and Q.

2. This was associated / in the USA / with changes in the environment.

3. However / only / for some Xs / this operation is defined.

4. From now on / since the two cases are almost identical / we will only refer to the first case.

5. With each operation / is associated / a number / which refers to the ranking.

6. At any time / it is possible to put / on hold / the application / by doing X.

7. It is thus possible / to select / from the database / only useful data.

8. Important parameters / are / both X and Y.

1. P and Q are particularly important factors that influence X.

2. In the USA this was associated with changes in the environment.

3. However this operation is only defined for some Xs.

4. Since the two cases are almost identical, from now on we will only refer to the first case.

5. A number is associated with each operation, which refers to the ranking.

6. It is possible to put the application on hold at any time by doing X. / By doing X, it is possible to put the application on hold at any time. / At any time, it is possible to put the application on hold by doing X.

7. It is thus possible to select only useful data from the database.

8. Both X and Y are important parameters.

2.8   shifting the parts of the phrase to achieve optimal order: 1

In each sentence below the cause is given first and then the effect. In which cases would it make more sense to mention the effect first rather than the cause?

1. Since it is the international language of research I study English.

2. On account of the fact that I am 2 m tall I have problems with air travel.

3. Due to our diet we don't have many overweight people in our region.

4. Owing to the fact that there was fog the plane was delayed by two hours.

5. As a consequence of the fact that my presentation was the last one of the day I decided to make it less formal and more fun.

6. A result of the rise in taxes is that spending has dropped dramatically.

These sentences make more sense to the reader by being rearranged as follows:

1. I study English since it is the international language of research.

2. I have problems with air travel on account of the fact that I am 2 m tall.

4. The plane was delayed by two hours owing to the fact that there was fog.

2.9   shifting the parts of the phrase to achieve optimal order: 2

Decide which you think is the best word order. If both are possible, what is the difference in meaning / emphasis?

(1a) In the second experiment, the plants accumulated lower amounts of selenium.

(1b) The plants accumulated lower amounts of selenium in the second experiment.

(2a) These findings highlighted in patients with severe disabilities a lack of cerebral activity.

(2b) These findings highlighted a lack of cerebral activity in patients with severe disabilities.

(3a) This will avoid discharging into the atmosphere in 2020 around eight million tons of debris.

(3b) This will avoid discharging around eight million tons of debris into the atmosphere in 2020.

(4a) This makes it possible to read the gear sensor with increased precision.

(4b) This makes it possible to read with increased precision the gear sensor.

(5a) In the following sections, X and Y are detailed.

(5b) X and Y are detailed in the following sections.

(6a) In this kind of study it is of primary importance to perform the analysis of standard compounds and reference materials, such as raw natural substances and / or natural materials artificially aged in the laboratory, to determine the specific ions and ion fragments.

(6b) In this kind of study in order to determine the specific ions and ion fragments, it is of primary importance to analyze standard compounds and reference materials, such as raw natural substances and / or natural materials artificially aged in the laboratory.

(7a) This occurs when in the original network there is a dependent voltage, as in the circuit of Fig 3b.

(7b) This occurs when there is a dependent voltage in the original network, as in the circuit of Fig 3b.

(8a) A peak was observed at m / z 426, relating to the molecular ion, along with peaks at m / z 411 and 408.

(8b) The presence of a peak at m / z 426, relating to the molecular ion, was observed along with peaks at m / z 411 and 408.

(1a) This order would probably be used when the author has just described the first experiment and now wants to alert the reader that the focus will now move to the second experiment

(1b) Here the focus is on the selenium rather than the experiment.

(2b), (3b), (4a) – These are the correct answers because they follow typical English word order of putting the direct object before the indirect object.

(5b) X and Y are the subject of the verb so it makes sense to put them at the beginning of the sentence. 5a is grammatical correct but would probably not be used by a native speaker.

(6b) This is best because it tells the reader the aim (i.e. determining the ions) near the beginning of the sentence rather than forcing the reader to wait for this information.

(7b) Direct object (voltage) before the indirect object (network)

(8a) This is best because the verb (observed) is close to the subject (peak)

28

2.10   shifting the parts of the phrase to achieve optimal order: 3

Rearrange and / or rewrite the sentences so that the information appears in a more logical order. Delete any redundancy.

1. A typical example is the following:

2. By eating too much of the wrong things, by drinking too much alcohol (especially wine and beer), and by not taking enough physical exercise, people may become overweight.

3. The table shows that for experimental purposes X is very useful.

4. In addition, in the mass spectrum are evident peaks at m / z 438, 411, and 410.

5. About the *TTC* output, it is computed by multiplying x by y.

1. The following is a typical example:

2. People may become overweight [for a number of reasons:] by eating too much …

3. The table shows that X is very useful for experimental purposes.

4. In addition, peaks at m / z 438, 411, and 410 are evident in the mass spectrum.

5. The *TCC* output is computed by multiplying x by y.

2.11 shifting the parts of the phrase to achieve optimal order: 4

Choose the sentence (a or b) that gives the information in the most reader-friendly order, i.e. that enables the reader to quickly assimilate the information contained in the sentence.

(1a) Control procedures: There are basically four procedures: up-shift, down-shift, neutral and start.

(1b) Control procedures: These procedures are basically four: namely up-shift, down-shift, neutral and start.

(2a) A known amount of the standard solution was added to the organic extracts, which were then stored in glass containers until their arrival at the lab.

(2b) The organic extracts were added with a known amount of the standard solution and stored in glass containers until their arrival at the lab.

(3a) These factors have, until now, been considered as irrelevant.

(3b) Until now these factors have been considered as irrelevant.

(4a) The presence of possibly undetected components was tested by Gas chromatography–mass spectrometry.

(4b) Gas chromatography–mass spectrometry was used to test for the presence of possibly undetected components.

(5a) Over the last 20 years, several exhibitions of the artists whose organization was carried out with the sponsorship of local banks have been set up.

(5b) Over the last 20 years, several exhibitions of the artists have been set up and organized with the sponsorship of local banks.

(6a) We present a method for screening, evaluating and comparing wood samples in a short time frame.

(6b) We present a method for screening, evaluating and the comparing in short time frame wood samples.

(7a) The most common approach is to analyze cross sections at different depths using optical and electron microscopy.

(7b) The most common approach is to analyze, using optical and electron microscopy, cross sections at different depths.

(1) a (2) a (3) b (4) b (5) b (6) a (7) a

2.12   reducing the number of commas and parts of the sentence

Complete the indented sentences so that they mean the same as Sentence A. Note that the word order in Sentence A is not correct.

(1a) In Fig. 2 a reference undeformed configuration, named X, and an adjacent deformed configuration, named Y, are shown.
(1b) Figure 2 ...

(2a) Ten datasets with the same X but a different Y, along with five datasets with a different X but a similar Y, were generated.
(2b) A number of datasets were generated: ...

(3a) The analytical steps, owing to the difficulties in measuring X, require some simplifications.
(3b) Due to ...

(4a) We can separate, by splitting these sections in the middle, P and Q.
(4b) By splitting ...

(5a) Concerning the role of education, it is given great emphasis in their society.
(5b) Their society ...
(5c) The role of ...
(5d) Great emphasis ...

(1b) Figure 2 shows ... named Y.

(2b) A number of datasets were generated: ten datasets with the same X but a different Y, and five with a different X but a similar Y.

(3b) Due to the difficulties in measuring X, the analytical steps require some simplifications.

(4b) By splitting these sections in the middle, we can separate P and Q.

(5b) Their society gives great emphasis to the role of education.

(5c) The role of education is given great emphasis in their society.

(5d) Great emphasis is given to the role of education in their society.

2.13   putting sentences into the correct order

Read this extract from a manual on good writing. Put the five sentences into the correct order

(a) For every author, there are hundreds or thousands of readers. It thus make sense for the author to spend an extra hour making a document readable, rather than forcing thousands of readers to spend an extra hour trying to understand the document.

(b) I get frustrated when I have to read a telephone number that is written like this: 00441618269987.

(c) 0044 161 826 9987 is an example of good readability and 'reader-centered' writing. The focus is on helping the reader to understand quickly and easily, even if it means the author having to make more effort.

(d) 00441618269987 is an example of poor readability. It is hard for the reader to assimilate. It is also an example of 'author centered' writing – the 'author' knows the number and is very familiar with it. He or she writes it down in the fastest way possible (thus not wasting his / her time) without thinking about how easy it will be for the reader to understand (and thus forces the reader to waste time).

(e) When I then dial the number I invariably make a mistake. It would be much easier to understand if it were written: 0044 161 826 9987.

(1) b

(2) e

(3) d

(4) c

(5) a

2.14 typical mistakes

Correct any mistakes in the following sentences. The mistakes are all related to word order.

1. This leaves intact for at least six weeks the sample.
2. Many are the substances that are harmful to human beings.
3. This book shares with the previous one several aspects.
4. For several years with this system we have had problems.
5. The female chimpanzees immediately after giving birth are generally quite weak.
6. After the written examinations each student has between May and June an oral exam.
7. The presence in written English of long sentences can cause problems for the reader.
8. In English is not commonly found a sentence construction that does not reflect the following order: subject verb object.
9. In the USA farmers have inadvertently introduced into the environment several dangerous species of insects.
10. One way to avoid such behavior in adults is treatment before the age of 16 with a high dose of insulin.

1. This leaves **the sample intact** for at least six weeks.
2. **There are many substances** that are harmful to human beings.
3. This book shares **several aspects** with the previous one.
4. We have had problems with this system for several years. / For several years we have had problems with this system.
5. **Immediately after giving birth** the female chimpanzees are generally quite weak / ... quite weak **immediately after giving birth**.
6. After the written examinations each student has **an oral exam** between May and June.
7. The presence **of long sentences** in written English can cause problems for the reader.
8. In English a sentence construction **is not commonly found** that does not reflect the following order: subject verb object.
9. **In the USA, farmers / Farmers in the USA** have inadvertently introduced several dangerous species of insects **into the environment**.
10. One way to avoid such behavior in adults is treatment **with a high dose of insulin** before the age of 16.

Section 3: Writing short sentences and paragraphs

TOPIC	ENGLISH FOR RESEARCH USAGE, STYLE, AND GRAMMAR	ENGLISH FOR WRITING RESEARCH PAPERS
Breaking up long sentences	15.1	Chapter 3
Writing paragraphs		Chapter 4
Highlighting findings through new paragraphs and short sentences		7.9, 8.1–8.4

There are basically four kinds of sentences.

Simple (subject + verb + object): *I finished my paper yesterday. I sent it to the journal.*

Compound (two simple sentences joined by a conjunction, e.g. and, because): *I finished my paper and I sent it to the journal.*

Complex (a simple sentence split by an intervening subordinate clause): *My paper, which had taken twomonths to write, was rejected by the journal.*

Compound + complex: *My paper, which had taken two months to write, was rejected by the journal because the referees said it made no contribution to the current state of the art. (29 words)*

The simpler the sentence, the easier it is for the reader to understand. This does not mean writing sentences containing only five or six words. But it does mean being aware that the last type of sentence (compound and complex) should not be used too often and should have a limit of generally not more than 30 words. If it is over 30 words, there is a chance that the reader will have to read it twice in order to fully understand it.

A. Wallwork, *English for Academic Research: Writing Exercises*,
DOI 10.1007/978-1-4614-4298-1_3, © Springer Science+Business Media New York 2013

3.1 dividing up long sentences: 1

Divide up these sentences into more manageable and shorter sentences that will help the reader understand the content better. You may need to rearrange the word order and / or delete unnecessary words.

Example
ORIGINAL The seeds, sterilised for 3 min. in NaOCl (1% available chlorine) and rinsed with distilled water, were germinated on moist filter paper (Whatman No. 2) in Petri dishes and grown in the dark at 23 °C till 72 hours.
REVISED The seeds were sterilised for 3 min. in NaOCl (1% available chlorine), and rinsed with distilled water. They were then germinated on moist filter paper (Whatman No. 2) in Petri dishes and grown in the dark at 23 °C.

1. Using automatic translation software (e.g. Google Translate, Babelfish, and Systran) can considerably ease the work of researchers when they need to translate documents thus saving them money (for example the fee they might have otherwise had to pay to a professional translator) and increasing the amount of time they have to spend in the laboratory rather than at the computer.

2. In order to establish a relationship between document length and level of bureaucracy in European countries and to confirm whether documents, such as reports regarding legislative and administrative issues, vary substantially in length from one language to another, we conducted an analysis of A, B and C.

3. The aim of our study was to assess changes in the level of tolerance of natives of one country towards immigrants over the course of a 50-year period in order to be able to advise governmental agencies on how to develop strategies based on those countries that have been more successful in reducing racism as already investigated in previous studies, but not in such a systematic way, and to establish correlations with data from the USA, which until now have been reported only sporadically.

4. Monolithic sorbent tip technology has proved to be efficient in removing interferences from copper and mercury salts, but it has not been tested for other materials and the recovery of proteinaceous material is often too small, giving rise to analysis problems.

5. Our results show that the performance of the system, in terms of throughput and delay, depends on several factors including the frame duration, the mechanisms for requesting uplink bandwidth, and the load partitioning, i.e. the way traffic is distributed, connections, and traffic sources within each connection.

1. Using automatic translation software (e.g. Google Translate, Babelfish, and Systran) can considerably ease the work of researchers when they need to translate documents. This can save them money, for example the fee they might have otherwise had to pay to a professional translator. It can also increase the amount of time they have to spend in the laboratory rather than at the PC.

2. We conducted an analysis of A, B and C. This was done in order to establish a relationship between document length and level of bureaucracy. We wanted to confirm whether or not documents, such as reports regarding legislative and administrative issues, vary substantially in length from one language to another.

3. We assessed changes in the level of tolerance of natives of one country towards immigrants over the course of a 50-year period. The main aim was to be able to advise governmental agencies on how to develop strategies based on those countries that have been more successful in reducing racism. This aspect has already investigated in previous studies, but not in such a systematic way. The second aim was to establish correlations with data from the USA, which until now have been reported only sporadically.

4. Monolithic sorbent tip technology C18 has proved to be efficient in removing interferences from copper and mercury salts, but it has not been tested for other materials. Moreover, the recovery of proteinaceous material is often too small, thus giving rise to analysis problems.

5. Our results show that the performance of the system, in terms of throughput and delay, depends on several factors including the frame duration, and the mechanisms for requesting uplink bandwidth. It also depends on the load partitioning, i.e. the way traffic is distributed, connections, and traffic sources within each connection.

3.2 dividing up long sentences: 2

The sentences below all come from the same Introduction. Divide up the sentences into more manageable and shorter sentences. You may need to rearrange the word order and / or delete unnecessary words.

1. The aim of this study was to assess the effects of sending children away to school at the age of eight (or earlier) and its impact on their adult life (particularly after the age of 50) and thus to reach some definitive conclusions as to whether boarding schools (i.e. those schools where children study and sleep) actually fulfill the important educational and social roles that they claim to have.

2. People who have attended boarding schools often have no realisation of the effect that leaving their parents at a very young age has had on their emotional development because the signs of this effect generally do not become sufficiently apparent until middle age and are often due to a kind of subconscious repression which is why such subjects do not make the connection between their current levels of over-emotiveness and their childhood lack of parental affection.

3. Questionnaires were sent to 5000 ex-boarding school adults with an age ranging between 40 and 60 all of whom had previously given permission to access their medical records and all of whom were or had been married, with the purpose of setting up a database of subjects' responses regarding their school time experiences and their experiences now as adults.

4. A substantial increase in sensitivity to emotional situations characterizes the first stages of adult life leading to a possible uncontrolled release of anger or apparently unexplained feelings of anxiousness that appear to come from nowhere and may last for several days thus making life quite difficult not only for the subjects themselves but also for those living around them.

5. Treatments for these subjects are often very expensive and technically difficult, and their effectiveness very much depends on the willingness of the subject to undergo therapy and on the degree of stress, emotional disturbance and marital discord that they had experienced.

1. The aim of this study was to assess the effects of sending children away to school at the age of eight (or earlier) and its impact on their adult **life, particularly** after the age of **50. The ultimate objective was** to reach some definitive conclusions as to whether boarding schools (i.e. those schools where children study and sleep) actually fulfill the important educational and social roles that they claim to have.

2. People who have attended boarding schools often have no realisation of the effect that leaving their parents at a very young age has had on their emotional **development. This is** because the signs of this effect generally do not become sufficiently apparent until middle age and are often due to a kind of subconscious **repression. In fact this delayed reaction** is why such subjects do not make the connection between their current levels of over-emotiveness and their childhood lack of parental affection.

3. Questionairres were sent to 5000 ex-boarding school adults with an age ranging between 40 and **60. All subjects** had previously given permission to access their medical records, and **all were** or had been **married. The aim was to set** up a database of subjects' responses regarding their school time experiences and their experiences now as adults.

4. A substantial increase in sensitivity to emotional situations characterizes the first stages of adult **life. This increase leads** to a possible uncontrolled release of anger or apparently unexplained feelings of **anxiousness. Such feelings** appear to come from nowhere and may last for several **days. Consequently life may be** made quite difficult not only for the subjects themselves but also for those living around them.

5. Treatments for these subjects are often very expensive and technically **difficult. Their** effectiveness very much depends on the willingness of the subject to undergo **therapy. The effectiveness also** depends on the degree of stress, emotional disturbance, and marital discord that they had experienced.

3.3　dividing up long paragraphs 1

The following extract is from an Introduction. It is one long paragraph and contains three very long sentences, averaging over 80 words each. Divide the paragraph into three shorter paragraphs, and break up each sentence into shorter more manageable sentences.

The aim of this paper is to confirm that how we speak and write generally reflects the way we think and that this is true not only at a personal but also at a national level, and to this end two European languages were analysed, English and French, to verify whether the structure of the language is reflected in the lifestyle of the respective nations. English is now the world's international language and is studied by more than a billion people in various parts of the world thus giving rise to an industry of English language textbooks and teachers, which explains why in so many schools and universities in countries where English is not the mother tongue, it is taught as the first foreign language in preference to, for example, Spanish or Chinese, which are two languages that have more native speakers than English. As a preliminary study, in an attempt to establish a relationship between document length and level of bureaucracy, we analysed the length of 50 European Union documents, written in seven of the official languages of the EU, to confirm whether documents, such as reports regarding legislative and administrative issues, vary substantially in length from one language to another, and whether this could be related, in some way, to the length of time typically needed to carry out daily administrative tasks in those countries (e.g. withdrawing money from a bank account, setting up bill payments with utility providers, understanding the clauses of an insurance contract). The results showed that...

The aim of this paper is to confirm that how we speak and write generally reflects the way we think, and that this is true not only at a personal but also at a national level.

Alternatively The two aims of this paper are firstly to confirm that how we speak and write generally reflects the way we think. And secondly, that this is true not only at a personal but also at a national level.

Two European languages were analysed, English and French, to verify whether the structure of the language is reflected in the lifestyle of the respective nations.

New paragraph English is now the world's international language and is studied by more than a billion people in various parts of the world.

This has given rise to an industry of English language textbooks and teachers. In fact, in many schools and universities in countries where English is not the mother tongue, it is taught as the first foreign language.

This choice is in preference to, for example, Spanish or Chinese, which are two languages that have more native speakers than English.

New paragraph As a preliminary study, we tried to establish a relationship between document length and level of bureaucracy.

We analysed the length of 50 European Union documents, written in seven of the official languages of the EU.

This was done to confirm whether documents, such as reports regarding legislative and administrative issues, vary substantially in length from one language to another.

We also wanted to know whether length of documents could be related, in some way, to the length of time typically needed to carry out daily administrative tasks in those countries.

These tasks included withdrawing money from a bank account, setting up bill payments with utility providers, understanding the clauses of an insurance contract. The results showed that …

3.4 dividing up long paragraphs 2

Divide this Review of the Literature into eight or more paragraphs.

A review of the literature in this field clearly shows that the majority of authors believe that there is an inherent difference between English and Latinate languages: English has been established as having a simpler and often more logical structure. Gestri et al. [2011], for example, contend that English is by nature a more synthetic language. Burgess [2001], in his seminal work on the subject, disputed some of Gestri's observations. Specifically, Burgess called into question the latter's GAS index, and eventually reformulated it into the SMOKEWARE index [2004]. Smith and Jones [2010] compared English and Spanish technical writing and found that English used about 30% less words to express the same concept. They confirmed previous research on the subject by concluding that English is an inherently simpler and more concise. A similar study was made by Ughi [2014] who reported on an interesting statistical analysis of typical phrases in the two languages. His findings were essentially the same as Smith's and Jones', but deviated in the percentages – 40% rather than 30%. Our work is a direct continuation of the work begun by Smith and Ughi, but with two essential differences: (1) The works quoted above make the unwarranted assumption that English has always been a simple language. We attempt to prove otherwise. (2) We go a step further than previous works, in that we establish a correlation between the simplicity of a language and the ease of life in the nation where that language is spoken. To prove these two points we developed a Verbosity Index, derived from 1000 recent scientific articles written in English, and the same number written in Italian (for full details see Sect. 4). The Verbosity Index was computed on the basis of the difficulty in comprehending an article, primarily in terms of sentence length – the higher the VI, the more difficult the understanding. The same process was then repeated for articles written 50 years ago. The results show that the English of 50 years ago has a comparable Verbosity Index to current French, Italian and Spanish, but is much higher than current English. Our findings demonstrate that English has become increasingly less verbose over the last 20 years. We believe that this trend can in part be attributed to such organizations as the Campaign for Plain English and Siegel and Gale (a company specialised in reducing the length and complexity of government documents). A concerted effort has been made to make English simpler. This has been done for two main reasons. Firstly, to make the written language more accessible to a wider variety of people (but not primarily those whose first language is not English). Secondly, for economic reasons it makes much more sense to have a document of one page rather than three. Not only is time saved in writing and reading the document, it also costs less to produce, and takes much less time to process, especially in the case of such documents as passport forms and tax declaration returns. We believe that our work has significant implications for all those countries whose citizens are habitually buried in bureaucratic procedures and forms.

Paragraph 2 begins: Smith and Jones

Paragraph 3 begins: Our work is

Paragraph 4 begins: (1) The works

Paragraph 5 begins: (2) We go a step

Paragraph 6 begins: To prove

Paragraph 7 begins: The same process

Paragraph 8 begins: Our findings

Paragraph 9 begins: A concerted effort

Paragraph 10 begins: We believe

42

3.5 dividing up long paragraphs 3

Divide up the following paragraph into three shorter paragraphs.

Most people tend to think of Lewis Carroll as some slightly eccentric character who wrote children's stories set in a wonderful make believe land that appealed both to kids and adults alike. But in fact his real name was the Reverend Charles Dodgson and he was far more than a writer. He was born in 1832 and spent much of his childhood doing magic shows for his brothers and sisters. He then went away to school at Rugby before getting his degree at Oxford University. His most famous books are Alice in Wonderland, written in 1865, and Through a Looking Glass which he wrote seven years later. Alice was based on the daughter of the Dean of Christ Church, which was the college at Oxford where Carroll later became Professor of Mathematics. He was in fact a terribly boring professor, so bad in fact that his students asked for him to be replaced. Besides writing children's stories and mathematical treatises, he also wrote an incredible number of letters. In fact from the age of 29 to his death in 1898, he wrote no less than 98,271 letters. Many of these letters were written in mirror language, or back to front, so that they had to be read from the end to the beginning, and most contained some kinds of puzzles. When he wasn't writing or listening to his musical box being played backwards Carroll invented all kinds of things including a prototype travelling chess set, double-sided sticky tape, and a new Proportional Representation scheme for electing members of parliament. In Carroll's system each candidate could give the votes given to him to another candidate. He might well have been the first person to make a self-photographing device and he later became one of the leading portrait takers of his time – notably of young girls like Alice.

Paragraph 2 begins: His most famous books

Paragraph 3 begins: When he wasn't writing

3.6 dividing up long paragraphs 4

Divide up the following paragraph into six shorter paragraphs.

Ask any non-native speaker (NNS) of English which language is simpler English, or their own mother tongue, and they will invariably say that their own language is more complex. Ask any native English teacher the same question, and they will say 'English'. So is the complexity of a language more linked to national pride or objective factors? It is interesting that complexity is also often considered as something positive rather than negative: the more something is difficult the better, in some inexplicable way, it is. English is simpler, NNSs say, because you say, for example, *I want, you want, we want, they want, I wanted, you wanted* etc., with only the third person in the present causing any irregularity. Whereas in many other languages each person has its own ending throughout the tense system. But is this a matter of complexity, or simply of memory? Once you've memorized the endings system of verbs in for example, Greek, Italian or German, there is little difficulty in their actual use. In English the opposite seems to be true – tenses are easy to form but their actual use is far more subtle. How many English teachers, let alone students, can explain clearly the difference between *I will do, I am doing, I am going to do, I will be doing, I am going to be doing, I will have done* and *I will have been doing*? Some of these future tenses don't even exist in most other languages. Another reason that NNSs use in their justification of English being simpler because it is written in nice short sentences. Actually, English is written in nice short sentences only because most people have made a conscious decision to write like that because they know their ideas will be communicated better if they do so. There are even organizations in Britain and the US whose revenues depend exclusively on making their clients' English clearer and more concise. But legal English, for example, is just as complicated as legal Arabic, Russian, Japanese or Hindi, perhaps because British and American lawyers want to hold on to their jobs just as much as their counterparts overseas – only they are the ones who should be able to interpret each other's deliberately contorted legalese. It is often claimed that the non-complex nature of English is also due to the fact that the same word, for instance *get*, is used in hundreds of ways. *get* in fact can be combined with almost every preposition and adverb to give another meaning, but many of these meanings can also be rendered with another, often Latin-derived (and therefore more formal) verb. English has an inbuilt capacity to be formal or informal depending on the choice of word – and English has a lot more words to choose from than the majority of other languages. But these examples of simplicity are often in NNS's eyes evidence of the inferiority of English; as indeed are the complexities, for example the near total lack of correspondence between the way a word is spelt in English and the way it is pronounced (e.g. though, one, business).

But this means that many other languages are unable to indulge in the word games the English are so particularly fond of: *The Beatles* for example is a mix of *beetles* (i.e. the insect) and *beat* (i.e. rhythm). A language should not be judged on how simple or complex it is, but on how expressive it is. English commands a vaster vocabulary and a far wider range of tenses than most other tongues. Other languages may have a much more flexible word order / sentence structure than English, and may even sound more eloquent and beautiful than English. At the end of the day, however, the average university-educated speaker, whether mother tongue English or not, knows about 25,000 words in their own language, and is equally capable as being as literate or as illiterate as they like.

Paragraph 2 begins: English is simpler, NNSs say,

Paragraph 3 begins: Another reason that NNSs use

Paragraph 4 begins: It is often claimed that

Paragraph 5 begins: But these examples of simplicity

Paragraph 6 begins: A language should not be judged

3.7 putting paragraphs into their most logical order

Put these four extracts from the Conclusions of a paper into a logical order.

(a) As international scientific English becomes the preserve increasingly of the non-native speaker, as it will do if English continues its path towards greater and greater dominance of the world of science, then the same thing will happen to scientific English as has happened to general Englishes. We will see the rise of varieties of scientific English or 'Scientific Englishes' which will deviate more or less from the standard.

(b) If English is now the preferred language of international communication in science, then within the sphere of scientific communication, English should be seen as what it is, the language of science, not the language of Englishmen or Americans.

(c) Where that style is a function of the needs of the genre and acts positively to communicate the science more effectively, then it should be respected. Where it is simply a matter of preferred native-speaker style, it can safely be ignored. The owners of international scientific English should be international scientists not Englishmen or Americans.

(d) In conclusion, therefore, I would argue that non-native speakers of English who communicate their science in English should not feel any sense of inferiority vis-a-vis the native speaker in this respect. The overriding framework in determining how to communicate should be the science itself, rather than the rhetorical style of the language.

(1) b

(2) a

(3) d

(4) c

3.8   writing short sentences: 1

Write five or six sentences about the research you are currently carrying out. If you have not started your research, imagine that you have. Write short sentences, as in the example below:

We investigated the meaning of life.
We used four different methodologies.
Each methodology gave contradictory results.
The results confirmed previous research indicating that we understand absolutely nothing.
Future research will investigate something more simple – the cerebral life of a PhD student.

3.9   writing short sentences: 2

Write five or six sentences on one or more of the following topics. Each sentence should contain a maximum of 20 words.

1. How has the role of women changed in your society in the last 50 years?

2. Should government funding of 'less practical' research areas, such as philosophy, history and theology, be cut?

3. All academic books should be free. Discuss.

4. Your ideal teacher / professor.

5. The downsides of the Internet.

3.10   writing short sentences: 3

Write five or six sentences on one or more of the following topics. Each sentence should contain a maximum of 15 words.

1. Daily life in 2075.

2. In what ways would a benign dictatorship be better than a democracy?

3. More should be spent on medical research than on defense.

4. Did the invention of the map or the clock cause the biggest change to human life?

5. What obligations do first world countries have towards third world countries?

Section 4: Link words: connecting phrases and sentences together

TOPIC	ENGLISH FOR RESEARCH USAGE, STYLE, AND GRAMMAR	ENGLISH FOR WRITING RESEARCH PAPERS
all link words	Section 13	4.14
also, in addition, besides, moreover	13.2, 13.3	3.7
although, even though, even if	13.4	
and, as well as, along with	13.5	3.5, 3.6
as far as x is concerned	13.1	
as, as it	13.6	
as, like	13.7	
because, since, as, in fact	13.8	3.9
both, and, either, or	13.9	
e.g., i.e., etc	13.10–13.11	
former, latter	13.13	
nevertheless, nonetheless, notwithstanding	13.14, 13.15	
owing due to, due to, as a result of, consequently, thus, etc	13.8	3.10, 15.12
thus, therefore, consequently, hence, so	13.18	
whereas, on the other hand, instead, although, however	13.14, 13.15, 13.17	3.8

A. Wallwork, *English for Academic Research: Writing Exercises*,
DOI 10.1007/978-1-4614-4298-1_4, © Springer Science+Business Media New York 2013

4.1   linking sentences and paragraphs

Analyze this extract, which is the beginning of a new section in a writing manual produced by NASA. How is each sentence linked to the previous one? The underlined words should help you to reveal the writer's structure.

Paragraph 1: The length of a sentence should generally not exceed about 35 words. Any sentence presents readers with the task of first identifying its constituent phrases, and then bearing them all in mind while their logical interrelation crystallizes. Forcing readers to bear in mind and fit together more than about 10 phrases * (each of about three words) is unnecessarily cruel. Your English instructor's joy at your ability to compose grammatically correct 200-word sentences must be disregarded as against your present goal of simplifying your reader's job.

Paragraph 2: An equally important rule is that a sentence should generally contain some indication as to how it is related to the preceding sentence or to the development of the paragraph.

*'phrases' is used here to mean a series of a clauses separated by commas, which together form one incredibly long sentence.

Paragraph 1, Sentences 1–3: A key word from one sentence is repeated into the next sentence, thus forming a simple chain of ideas that the reader can easily follow.

Paragraph 1, Sentence 4: This is the final sentence in the first paragraph. It recalls all the key words from the whole paragraph and acts as a mini summary.

Paragraph 2, Sentence 1: This connects back to the rule given in Paragraph 1.

4.2   deleting unnecessary link words

Note how in the previous exercise the author did not use any link words in his paragraph. Instead, ideas are connected together by a logical progression of ideas. Look at the same extract, which this time includes link words. Do they add anything to facilitate your understanding of the text, or do they just distract you?

It is worthwhile noting that the length of a sentence should generally not exceed about 35 words. ***In fact,*** any sentence presents readers with the task of first identifying its constituent phrases, and then, ***as a consequence,*** bearing them all in mind while their logical interrelation crystallizes. ***In addition,*** forcing readers to bear in mind and fit together more than about 10 phrases (each of about three words) is unnecessarily cruel. ***Moreover, as far as technical writing is concerned,*** your English instructor's joy at your ability to compose grammatically correct 200-word sentences must be disregarded as against your present goal of simplifying your reader's job.

Of course, it goes without saying that an equally important rule is that a sentence should generally contain some indication as to how it is related to the preceding sentence or to the development of the paragraph.

4.3 deciding when link words are necessary

The link words used in the previous exercise are all redundant. This is not always the case, as you can see from another paragraph written by the same author. Decide what role these three link words play.

Different writers have different methods of organizing their reports, and some seem to have no discernible method at all. Most of the better writers, **however**, appear to be in remarkably close agreement as to the general approach to organization. This approach consists of stating the problem, describing the method of attack, developing the results, discussing the results, and summarizing the conclusions. You may feel that this type of organization is obvious, logical, and natural. **Nevertheless**, it is not universally accepted. **For example**, many writers present results and conclusions near the beginning, and describe the derivation of these results in subsequent sections.

However – alerts the reader that what the author is saying now contrasts with the idea presented in the previous sentence.

Nevertheless – is placed prominently at the beginning of the sentence. It catches the reader's eye and alerts him / her that a new viewpoint is about to be put forward.

For example – this is used to give an instance of what has been said before.

These link words all serve to show how each sentence relates to what has been said before. Without these link words, the reader would be forced to figure out the author's train of thought. However, the author only uses link words when they really serve a purpose.

4.4 choosing best link word

In the following Abstract where possible choose (c). Where not possible choose the most appropriate linker (a or b).

Dreams play a key role in our lives. **(1)** *(a) Also, they (b) They also (c) They* reveal important patterns in the way our brains work. **(2)** *(a) Therefore it is (b) It is thus (c) It is* important to understand how they are created, how they relate to each other and what their effects on our waking lives are.

This work presents the design, realization and validation of a new paradigm to effectively teach dream mechanics in secondary schools. **(3)** *(a) In particular the (b) Specifically, the (c) The* project was devised in close collaboration with high school teachers and offered to 180 students. The guiding principle was that classes should have a relevant laboratory component, where students could have an active role with hands-on training. **(4)** *(a) At the same time, students (b) In fact, students (c) Students would* have the means to learn the mathematical tools and the underlying physical models, used to describe experiments, analyze data and discuss the results. **(5)** *(a) Last, but not least, teachers (b) Lastly teachers (c) Teachers* should be able to motivate students to study dream mechanics, by highlighting its connections to everyday life and to other disciplines such as anthropology, philosophy and medicine.

(6) *(a) Bearing in mind all these objectives, we (b) We thus (c) We* formulated the following learning program based on lectures, seminars, experiments, data analysis and discussion of the results. The workshops were held in the classroom, and the theoretical framework and software tools were introduced. We started with the study of nightmares using an EMR device. **(7)** *(a) In addition practical (b) Furthermore practical (c) Practical* examples of nightmare mechanics were investigated both with a brain scan and a SOGNO circuit. **(8)** *(a) In actual fact we (b) In detail we (c) We* investigated dream mechanics in different cultures: **(9)** *(a) specifically, Australian (b) namely, Australian (c) Australian* aborigines, the Italian mafia and Japanese industrialists. **(10)** *(a) Finally, the (b) In summary the (c) The* interaction of dream scenarios with everyday reality was investigated.

(1) c (a and b also possible)

(2) b or a (c also possible)

(3) c

(4) a

(5) b (a is too informal, c also possible)

(6) b (a is too long, c also possible)

(7) c (a and b also possible)

(8) c

(9) c (the preceding colon makes a and b redundant, without the colon a and b would be fine)

(10) a

4.5   reducing the length of link words / phrases

Below is a list of phrases that are typically used at the beginning of a sentence in order to link it in some way to the previous sentence. The phrases could all be replaced with one of the following words: a) Note that b) Clearly c) In fact d) Since

Decide which words (a, b, c, or d) could replace the phrases below.

1. It is worthwhile noting that …,

2. As a matter of fact …,

3. It must be emphasised that …

4. It is interesting to observe that …

5. It is worthwhile bearing in mind that …

6. As the reader will no doubt be aware …

7. Obviously, this does not mean that …

8. It may be noticed that …

9. It is possible to observe that …

10. On the basis of the fact that …

(1) a	(6) b
(2) c	(7) b
(3) a	(8) a
(4) a	(9) a
(5) a	(10) d

4.6 shifting the position of link words expressing consequences

Replace the phrases in italics with thus. *Relocate* thus *to a different position and make any other necessary changes. The idea is that where possible link words should be placed later in the sentence rather than at the beginning.*

Example: *On account of this*, we decided to repeat the experiments.
= We **thus** decided to repeat the experiments.

1. *For this reason*, firms offering such goods need to make more effort in order to be aware of competitors.

2. *This means that* the diffusion of such processes is likely to increase the total number of errors made.

3. A licence agreement may be renewed after ten years. *In other words*, this makes software protection potentially infinite.

4. *As a consequence*, companies are devoting more and more attention to managing their brands.

5. *In line with this reasoning*, such firms should have a higher incentive to deposit trademarks.

1. Firms offering goods **thus** need to ...

2. The diffusion of such processes is **thus** likely ...

3. A licence agreement may be renewed after ten years **thus** making software ...

4. Companies are **thus** devoting ...

5. Such firms should **thus** have ...

54

4.7 using link words to give additional neutral information

Connect or combine the sentence in the first column with the sentence in the second column. Make any changes that you think are appropriate. Where necessary, one of the following link words: and, for example, e.g., such as, in addition, consequently

Example (see 1 below): Metallurgists study metals, how they can be extracted, and what properties they hold. **In addition**, they investigate how the properties can be modified in order to produce alloys

	TOPIC	ADDITION / EXAMPLE
1	Metallurgists study metals, how they can be extracted, and what properties they hold	Metallurgists investigate how the properties can be modified in order to produce alloys
2	Latinate languages are those languages that are based on Latin	French, Italian, Romanian, Portuguese, Spanish
3	Platinum is a rare metal	Platinum is extremely expensive
4	Synthesizers are a type of musical instrument that uses circuits to produce simple waveforms	Such circuits are called oscillators
5	Many of the first immigrants into what is now the USA were from Great Britain. The USA subsequently became the richest country on the planet	English is now the international language of commerce and science
6	Some types of chemical reaction take place rapidly, others very slowly	Explosions are examples of rapid reactions, rusting is an example of a slow reaction

1. ... properties they hold. **In addition**, they investigate

2. ... are based on Latin, **for example** French ...

3. Platinum is a rare metal **and** is extremely expensive.

4. Synthesizers are a type of musical instrument that uses circuits, **called oscillators**, to produce simple waveforms.

5. ... on the planet. **Consequently**, English is

6. Some types of chemical reaction take place rapidly, **such as explosions / (e.g. explosions)**, others very slowly, **such as rusting / (e.g. rusting)**.

4.8 using link words to give additional positive information

Combine the information in the first column with the additions in the second column. You can either make a longer sentence or create an additional new sentence. By combining these sentences you should be able to produce an Abstract on the topic of greening the Internet. Note: many of the sentences in the box are in note form and will need to be modified and / or expanded. You may find that the link words in column three help you to think of ways to combine the information, but you are not obliged to use them.

	INITIAL STATEMENT	ADDITION	LINK WORD
1	Greening the Internet: subject of research since the late 2000s.	greening = ways to reduce 8% of the world's energy that is currently consumed by Information and Communication Technologies (ICTs).	*i.e.*
2	Performance of Internet-based architectures no longer in terms of speed of computation.	Now based on power consumption.	*rather than*
3	Our method: shared power sources, advanced cloud computing.	Avoids drawbacks of other methods (e.g. x and y).	*thus*
4	Innovative method: reduces power consumption of typical industrial user by 25%.	Household user's PC power consumption reduced by 10%.	*respectively*
5	Our results show: Tokyo global warming limits can be met	If methodology put into practice: Antarctic ice melting will be delayed by further 20 years.	*in fact*
6	Applications of our findings: other industrial equipment and household appliances.	Future research: investigate how to reach 50% reduction in ICT-related power consumption in industry.	—

Greening the Internet, **i.e.** finding ways to reduce the 8% of world's energy currently consumed by Information and Communication Technologies (ICTs), has been the subject of research since the late 2000s. The performance of Internet-based architectures is now based on power consumption **rather than** speed of computation. Our method exploits shared power sources and advanced cloud computing, and **thus** avoids the drawbacks of other methods (e.g. x and y). Our innovative method reduces the power consumption of a typical industrial and household user by 25% and 10%, **respectively**. Our results show that the Tokyo global warming limits can be met. **In fact**, if our methodology is put into practice, Antarctic ice melting will be delayed by a further 20 years. Our findings could be applied to other industrial equipment and household appliances. Future research will investigate how to reach a 50% reduction in ICT-related power consumption in industry.

4.9 using link words to give additional negative information

Combine the sentences in the first column with additions in the second column, either to make a longer sentence or to create an additional new sentence. This should then create a paragraph that describes the limitations of the current literature on this topic. You may find that the link words in column three help you to think of ways to combine the information, but you are not obliged to use them.

	INITIAL STATEMENT	ADDITION	LINK WORD
1	To the best of our knowledge, no other authors have ever studied the relationship between the level of bureaucracy in a country and the complexity of the language spoken in that country.	The effect of this level of bureaucracy on the people of that country has not been investigated.	*nor*
2	Langue *et al* (2013) only investigated two countries: Spain and France.	Langue did not investigate Spanish or French ex-colonies (i.e. where Spanish and French are spoken).	*and moreover*
3	Their study failed to take into account that levels of bureaucracy are not uniform throughout the same nation.	Their study only considered level of bureaucracy in the national health system, not in local or national government in general.	*in fact … but*
4	A major limitation of their method is that it only exploits a small sample basis.	Sampling was only carried out once – thus it cannot account for changes that may take place over time.	*in addition*
5	Their conclusion regarding a lack of relationship between bureaucracy and complexity of language is erroneous because they only investigated two languages.	The two languages are very similar – France and Spain are neighboring countries with very similar languages.	*finally … in any case*

To the best of our knowledge, no other authors have ever studied the relationship between the level of bureaucracy in a country and the complexity of the language spoken in that country. **Nor** has the effect of this level of bureaucracy on the people of that country been investigated. Langue *et al* (2013) only investigated two countries: Spain and France, **and moreover** they did not study Spanish or French ex-colonies (i.e. where Spanish and French are spoken). Their study failed to take into account that levels of bureaucracy are not uniform throughout the same nation. **In fact**, they only considered the level of bureaucracy in the national health system, **but** not in local or national government in general. A major limitation of their method is that it only exploits a small sample basis. **In addition**, sampling was only carried out once thus failing to account for changes that may take place over time. **Finally**, their conclusion regarding a lack of relationship between bureaucracy and complexity of language is erroneous because they only investigated two languages, **which are in any case** very similar – France and Spain are neighboring countries with very similar languages.

4.10   making contrasts

Connect each statement with its contrasting statement using one of the following link words: on the other hand, on the contrary, in contrast, in reality, but, instead, whereas, however, nevertheless, despite this. *Note: you can choose to use the same link word more than once.*

	STATEMENT	CONTRAST
1	There is very strong evidence that the universe began around 13.7 billion years ago.	Some religious fundamentalists believe that the world was created on October 22, 4004 BC.
2	Satellites can record data on snow melting, earthquake movements, cutting down of rainforests, the health of crops etc.	Scientists are needed in order to interpret the data.
3	The Celts were a dominant power in Europe around 50 BPE [*before present era*] and traded with the Greeks and Romans.	Greek and Roman culture had little influence on the Celtic civilization.
4	Centipedes are hunters.	Millipedes are herbivores.
5	Traditionally it was thought that men don't express their feelings or talk about their problems because they are repressed.	Men are able to talk about their personal problems but do not see the utility of doing so.
6	When water boils in an uncovered recipient it produces about 2,000 times its own volume of steam.	In a sealed recipient, steam cannot expand and steam pressure increases
7	English tends to be written in reader-centered style, in a way that makes the topic as easy as possible to understand for the reader.	Many other languages are author-centered and the reader has to decipher the meaning.
8	Current thinking claims that fructose is not harmful to the health.	Fructose is a poison. We prove that it can be classified as deleterious to the health in the same way as tobbaco, and ethanol in alcohol.

1. … billion years ago. **Despite this,** some religious …

2. … crops etc. **(No link word required)** Scientists are …

3. … greeks and Romans. **However,** Greek and …

4. Centipedes are hunters, **whereas** millipedes are herbivores.

5. **…** they are repressed. **In reality,** men are able …

6. … of steam. **In contrast / On the other hand**, in a sealed …

7. **…** for the reader. Instead / . **In contrast / On the other hand** many other languages …

8. … to the health. **On the contrary / In reality**, fructose is …

4.11   making evaluations

Write one advantage (pro) and one disadvantage (con) for each of the topics below. Then connect your two phrases together.
Useful link words: although, though, despite the fact that, nevertheless, on the other hand, but, instead

Example: Facebook
PRO: great way to keep in contact with friends
CON: people can upload embarrassing photos of you and say embarrassing things about you
CONNECTED SENTENCES: Although Facebook has become the most popular means of keeping in contact with friends, it does have one major downside: users have little control of what other people say about them or on the photos that can be uploaded.

1. cell phones
2. democracy
3. doing a Master's
4. electric cars
5. genetically modified organisms
6. marriage
7. space research
8. studying philosophy
9. travelling abroad
10. vegetarianism

4.12 connecting sentences by repetition of key word or a derivation of the key word

Read the text below and note how each sentence is connected to the next.

Nobody can predict where and when a ***thunderstorm*** will form. But (1) *it* starts with the formation of a normal cloud, which is formed from ***water vapor***. (2) ***This vapor*** accumulates by evaporation from plants, from evaporation from lakes and streams and surface water. The (3) ***water vapor*** then rises in the atmosphere in thermals in the boundary layer where it ***condenses***. (4) ***This condensation*** is the first stage of cloud formation.

How sentences are connected to each other: (1) *it* refers back to previous noun, (2) same word repeated, (3) more specific version of same word mentioned at beginning of previous sentence, (4) different form of root word.

Complete the following sentences, by repeating in some form the word in bold. Each _____ represents a separate word.

1. Pompeii was **discovered** in the 18th century. _____ _____ led to a revolution in Western design. The rich colors ...

2. This may be **possible** in the future. However such a _____ rests with governments.

3. Editors and referees often **reject** papers. _____ _____ are often the cause for anger or dismay, particularly amongst young researchers.

4. The engine **performed** extremely well. In fact _____ _____ was much better than had been expected.

5. Airports have a series of **restrictions** on what passengers can take on board. _____ _____ will soon be lifted.

6. They **know** how to solve such problems. _____ _____ is extremely important when ...

7. The petroleum industry has **suppressed** evolutions in water-powered cars. _____ _____ means that cars are much more expensive to run than they need to be.

8. They **forecast** increases in pollution levels back in the 1950s. _____ _____ were, however, largely ignored until ...

(1) This discovery

(2) possibility

(3) Such rejections

(4) its performance

(5) These / Such restrictions

(6) Such / This / Their knowledge

(7) This suppression

(8) These / Such forecasts

4.13   describing processes

Describe one *of the processes (1–5) below using what you consider to be the relevant steps (a-f).*

(a) *Formally define the process.*

(b) *Briefly outline the purpose or function of the process.*

(c) *Describe any recent developments or refinements in the process.*

(d) *Describe the first step in the process – what is it? what is its purpose? why is it performed?*

(e) *Describe subsequent steps as above.*

(f) *Draw some conclusions: advantages, limitations, specific uses.*

In each case refer the process to how it is typically followed in your country.

Useful link words: first, firstly, initially, at the beginning, secondly, thirdly, after this, before this, next, subsequently, then

1. The education system from the age of 13 up to and including graduation from university.

2. The election system.

3. Setting up a bank account.

4. Buying a plane ticket online.

5. Buying a cell phone and getting a cell phone number.

4.14 describing causes

Write a possible cause for three or more of the following. Useful link words: because, due to, as a result of, consequently, thus, this means that

1. Vandalism in inner-city areas.

2. The low birth rate in Western Europe.

3. Fewer people attending church in the West.

4. Children whose only ambition is to become famous.

5. The lower attention span of young people today compared to 30 years ago.

6. The number of English words (20,000+) adopted into Japanese.

7. The rejection by editors and referees of many scientific papers.

8. The increase in the number of e-books.

9. The massive success of social networks.

10. The constant increase in the number of blogs.

4.15 describing effects and consequences

Write an effect / consequence for the following.
Useful link words: thus, consequently, therefore, so

1. Footballers being paid huge quantities of money.

2. Corruption amongst politicians.

3. Banks lending money to those who cannot afford to repay it.

4. Countries relying on nuclear power for their energy.

5. China buying up huge areas of land in Africa.

6. US foreign policy.

4.16 making contrasts, concessions, qualifications, reservations, rejections

Write one or two sentences combining the information given in the three columns in the table.

Useful link words: although, though, but, despite this, however, nevertheless, which

	POSITIVE / NEUTRAL POINT	NEGATIVE POINT	DESPITE THIS
1	English is the language of research.	Difficult to understand the spoken language for non natives.	Relatively easy to learn compared to Chinese, Russian, German.
2	Manuscripts are subjected to peer review, which is a vital process.	May delay publication of important results for six months or more.	Ensures good quality papers with reliable and replicable results and methodologies.
3	Conferences enable researchers to exchange ideas.	Considerable expense of reaching conference location + cost of accommodation.	Researchers can make good contacts that may lead to collaborations and / or funding.
4	There is an abundance of Master's courses.	High cost + often only used to gain extra points in order to obtain research position.	Number and type of courses increasing every year.
5	Free online journals.	Author often has to pay a fee to the journal.	Rapid publication of results. Easily accessible.

1. English is the language of research, **though** it is difficult to understand the spoken language for non natives. **However**, it is relatively easy to learn compared to Chinese, Russian, German.

2. **Although** peer review may delay the publication of important results for six months or more, it **nevertheless** ensures good quality papers with reliable and replicable results and methodologies.

3. Conferences enable researchers to exchange ideas **but** at considerable expense (e.g. the cost of reaching conference location and the accommodation). **However**, researchers can make good contacts that may lead to collaborations and / or funding.

4. There is an abundance of Master's courses, **which** are very expensive and are often only used to gain extra points in order to obtain research positions. **Despite this**, the number and type of courses is increasing every year.

5. Free online journals enable the rapid publication of results and are easily accessible. **However**, the author often has to pay a fee to the journal.

4.17 outlining solutions to problems

Write short paragraphs which combine the information given in the three columns in the table. Only use link words where necessary.

	PROBLEM	CONSEQUENCE	SOLUTION
1	The waste (e.g. plastic bags, old TV sets) of developed countries often ends up in developing countries.	Children in developing countries often involved in sifting through waste. Pollutes the environment of the developing country.	Developed countries deal with their own waste by: • consuming less • using less packaging • higher% recycling
2	Only the relatively rich have access to the Internet.	The poor miss out on: • news about their own country • job opportunities • cheaper products and services	Free (i.e. no cost) Internet for low income families. Free courses on PC and Internet use.
3	Brain drain (top scientists leaving poorer country for richer country)	The poorer country loses its best scientists and thus revenue sources. Gap between poor and rich countries increases.	Rich countries set up and fund labs in the poorer country, without 'stealing' that country's scientists.
4	Most presentations at international conferences are poorly presented, poorly structured, and boring.	The value of much research is lost.	Courses on how to give interesting and effective presentations should be held in all institutes of the world, and should be funded by richer nations.
5	The majority of research into medicine is aimed at treating illnesses that are prevalent in the industrialised world.	Diseases that affect huge areas of Africa are neglected.	Priority should be given to the numbers of people affected by a disease, rather than the geographical location of the sufferers.

1. The waste (e.g. plastic bags, old TV sets) of developed countries often ends up in developing countries with the result that children in developing countries are often involved in sifting through waste. Moreover, this waste pollutes the environment of the developing country. In order to avoid such problems, developed countries should deal with their own waste by consuming less, using less packaging, and recycling more.

2. Only the relatively rich have access to the Internet. This means that the poor miss out on news about their own country, job opportunities and finding cheaper products and services. A solution to this problem would be free (i.e. no cost) Internet for low income families. In addition, there could be free courses on PC and Internet use.

3. Top scientists in the underdeveloped world often leave their poor country for a richer country – this is known as the brain drain. The poorer country loses its best scientists and thus revenue sources. Consequently, the gap between poor and rich countries increases. Rich countries could help to ameliorate this brain drain by setting up and funding labs in the poorer country, without 'stealing' that country's scientists.

4. Most presentations at international conferences are poorly presented, poorly structured, and boring. This means that the value of much research is lost. To counteract such problems, courses on how to give interesting and effective presentations should be held in all institutes of the world, and should be funded by richer nations.

5. The majority of research into medicine is aimed at treating illnesses that are prevalent in the industrialised world. However, diseases that affect huge areas of Africa are neglected. Priority should be given to the numbers of people affected by a disease, rather than the geographical location of the sufferers.

4.18   outlining a time sequence

Look at the advances in telephony below, which are divided up into three groups of years (A, B, C). Look at the example for Group A, then write something similar for Groups B and C. For Group C imagine that the year is now 2030.

Example
Group A

1876	First words spoken on a telephone
1877	First phone sold
1880	London's first phone directory (255 names)
1889	First coin pay phone
1910	Around 10 million people have a telephone

In 1876 the first words were spoken on a telephone and a year later the first telephone was sold. By 1880 there were already 255 names listed in London's first telephone directory. Within the next decade, around 10 million people had acquired a telephone.

Group B

1983	First cordless phone sold
1984	First videoconferencing system sold
1999	First full internet service on cell phones
2003	First Skype call made
2006	100 million people use Skype

Group C

2020	Number of mobile phone users one hundred times that of fixed phone users
2020	Globally, more women than men now their own cell phone
2021	First telephone to automatically translate from one language into another while speakers are speaking

Group B: The first cordless phone was sold in 1983, and the first videoconferencing system the following year. It was not for another 15 years before cell phones had Internet access. In 2002 the first Skype call was made and within only three years, 100 million people had subscribed to the service.

Group C: In 2020, the number of mobile phone users reached one hundred times that of fixed phone users. At the same time, the number women superseded the number of men owning a cell phone. A year later saw the introduction of the first telephone to ...

4.19   explaining figures and tables: making comparisons

The table below shows the results of a survey on PhD students. The students were asked what problems they had in writing research papers. They were NOT given the items in the table below, instead they decided the answer by themselves without prompting. Write a few simple sentences comparing what the students reported.

Useful link words: in fact, on the other hand, whereas, while

Example

The majority of students found the results and discussion sections to be the most difficult to write. In fact over four times as many students found these sections more difficult to write than the Abstract.

QUESTION	MOST COMMON ANSWERS	LEAST COMMON ANSWERS
Which section of a paper do you find the most difficult to write?	Results and Discussion 63%	Methodology / Materials 2%
	Abstract 15%	Bibliography 1%
	Introduction and Review of the Literature 8%	don't know 3%
	Conclusions 8%	
What aspects of your grammar are worried about?	articles (*the, a*) 25%	prepositions 12%
	tenses 20%	phrasal verbs 10%
	relative clauses (*that* vs. *which*) 15%	conditionals 9%
	other 18%	active vs. passive 9%
Which is the most relevant reason for a referee to reject a paper in terms of the quality of English?	poor vocabulary 24%	spelling mistakes 4%
	redundancy / lack of conciseness 22%	plagiarism 4%
	grammar errors 22%	contribution not clear due to poor language skills 2%
	poor structure of overall paper 11%	
	poor sentence and paragraph structure 11%	

4.20   making evaluations and drawing conclusions: 1

In one or two paragraphs, summarize the findings of the survey in the previous exercise by interpreting and elaborating on the students' answers. Conclude by outlining your own opinion and contrasting it with the results of the survey.

4.21   making evaluations and drawing conclusions: 2

Choose three of the inventions below. Write an evaluation of the inventions in terms of: a) feasibility, b) cost, c) utility, d) likelihood of widespread acceptance and usage if the invention was feasible and its cost not prohibitive.

1. A pill, given at birth, that can automatically treat any hereditary disease.

2. A car that runs on oxygen.

3. X-ray glasses.

4. A cream to cover the human body to render it invisible.

5. A teletransporter.

6. A software program that analyses your methodology and results, compares them with previous findings, highlights possible limitations, suggests applications, and then automatically writes your paper for you.

7. A microphone which at a conference automatically translates what you are saying in your mother tongue into perfect English.

Section 5: Being concise and removing redundancy

TOPIC	ENGLISH FOR RESEARCH USAGE, STYLE, AND GRAMMAR	ENGLISH FOR WRITING RESEARCH PAPERS
redundancy and writing less	15.4	5.1, 5.2
cutting individual words		5.3
using verbs rather than nouns	15.5	5.4, 5.5
link words		5.6
choosing short words and expressions		5.7, 5.8, 5.9
avoiding impersonal expressions	16.3.3	5.11
figures and tables	27.1	5.13
titles		11.11
review of the literature		15.7
methods / experimental		20.7

Being concise is essential in scientific writing. It helps your findings stand out and also enables the reader to find and absorb information quickly. There are many ways to revise your writing and make it more concise:

- Deleting words, phrases and sentences, but without making any other further changes.
- Replacing a phrase with a single word, again without making other changes.
- Replacing verbs with nouns.
- Rewriting.

While doing the exercises in this section, try to imagine that you have been given strict instructions by the editor or conference organizers to reduce your text to the minimum. In the key to the exercises you may consider some of the deletions to be quite extreme, but they are designed to show you what kinds of deletions are possible.

Note that the English in all sentences is correct. The problem is just that they are full of redundancy.

A. Wallwork, *English for Academic Research: Writing Exercises*,
DOI 10.1007/978-1-4614-4298-1_5, © Springer Science+Business Media New York 2013

5.1 removing individual redundant words

In each of the sentences below there is one redundant word. Find the redundant word and delete it, as in first sentence.

1. One ~~suitable~~ method is to separate the men from the women.

2. Dynamism and velocity are typical characteristics of this species.

3. Their head office is located in London.

4. If there is water present in the system, this may cause rust.

5. The results obtained highlight that $x=y$.

6. We have made an advance plan for the project.

7. We have seen from actual experience that this solution is not effective.

8. Another possible approach is to calculate P before Q.

9. Chemical reactions between organic materials and pigments lead to discoloration phenomena.

10. Our research activity consists of x and y.

1.	suitable	6.	advance
2.	characteristics	7.	actual
3.	located	8.	possible
4.	present	9.	phenomena
5.	obtained	10.	activity

5.2   removing several redundant words: 1

Each of the sentences below contains words or phrases that can be deleted without requiring any other changes. Delete such words and phrases.

The solution ~~adopted~~ was to ~~carry out a~~ test ~~of~~ all the software on the market.

1. As we have already noted in Section 4.2.1, the presence of x can influence y.

2. However, we have to make use of other techniques.

3. Paint samples, as described previously, normally contain mixtures of different substances.

4. In comparative terms, there is no real difference between x and y.

5. Also, we present simulation results that will provide a twofold contribution: (1) confirm the effectiveness of …, and (2) highlight the importance of …

6. The document was written in the English language and the contents represent a new innovation in the sector of telecommunications.

7. In our documents, we only use the Track Changes functionality for revisions.

8. Rows and also cells are highlighted in different colors in order to give a more effective view.

9. We identified a number of key factors that might affect, at least in principle, the performance.

10. They have shown that we should also consider the possibility of doing the testing in advance.

11. It is important to note that one plus one is equal to two.

12. In general, it has been established that the phenomenon of e-commerce is often characterized by the absence of a direct contact between seller and buyer [Mugandi, 2016].

13. In particular, the scope of this section is to show the strengths of …

14. Most of the organic materials that can be encountered in a painting are characterised by a macromolecular nature.

15. The use of this method is recommended in all cases where x=y.

1. As ~~we have already~~ noted in Section 4.2.1, the presence of x can influence y.

2. However, we have to ~~make~~ use ~~of~~ other techniques.

3. Paint samples~~, as described previously,~~ normally contain mixtures of different substances.

4. ~~In comparative terms,~~ there is no real difference between x and y.

5. Also, we present simulation results that ~~will provide a twofold contribution~~: (1) confirm ..

6. The document was written in ~~the~~ English ~~language~~ and the contents represent a new innovation in ~~the sector of~~ telecommunications.

7. [In our documents,] we only use ~~the~~ Track Changes ~~functionality~~ for revisions.

8. Rows and ~~also~~ cells are highlighted in different colors ~~in order~~ to give a more effective view.

9. We identified ~~a number of~~ key factors that might affect~~, at least in principle,~~ the performance.

10. They have shown [that] we should ~~also~~ consider ~~the possibility of doing the~~ testing in advance.

11. ~~It is important to~~ note that one plus one is equal to two.

12. ~~In general, the phenomenon of~~ e-commerce is often characterized by the absence of a direct contact between seller and buyer [Mugandi, 2016].

13. ~~In particular,~~ the scope of this section is to show the strengths of …

14. Most ~~of the~~ organic materials ~~that can be encountered~~ in a painting are ~~characterised by a~~ macromolecular ~~nature~~.

15. ~~The use of~~ this method is recommended in all cases where x = y.

5.3   removing several redundant words: 2

The sentences below all make references to the literature. Delete any unnecessary words or phrases. You do not need to make any other changes.

~~It has been shown that~~ microwaves significantly shorten reaction times, giving good reaction yields and a reduced decomposition of labile sugars [14–16].
Smith and Jones proved that x is greater than y. ~~In particular,~~ they demonstrated that

1. In this paper we give an overview of the most relevant existing work in this area of research.

2. It is in fact known that glucose is one of the products of photosynthesis in plants [Yang, 2014].

3. Experience teaches us that this is generally the best approach (see our previous papers: 12, 22, and 34).

4. Several authors in the past and also more recently have shown that substantial improvements in performance can …

5. Several algorithms have been proposed during the last few years and their properties have been thoroughly analyzed in the literature [3].

6. It should be borne in mind that there may be a high environmental contamination level [Smrkolj, 2005].

7. Finally, it is known that in archaeological waxes, hydrocarbons sublimate over time and it cannot be excluded that a similar phenomenon may occur in a thin film of oil [Bakali, 2014].

8. Another proposal is presented in [12], where the authors use a time-series analysis with travel speed simulation to predict future trajectories. In particular, they use a process based on range querying with spatial-temporal constraints on moving object database.

1. In this paper we ~~give an~~ overview ~~of~~ the most relevant existing work in this area of research.

2. ~~It is~~ in fact ~~known that~~ glucose is one of the products of photosynthesis in plants [Yang, 2014].

3. ~~Experience teaches us that~~ this is generally the best approach (see our previous papers: 12, 22, and 34).

4. Several authors ~~in the past and also more recently~~ have shown that substantial improvements in performance can ...

5. Several algorithms have been proposed ~~during the last few years~~ and their properties have been thoroughly analyzed ~~in the literature~~ [3]. / Several algorithms have been proposed ~~during the last few years~~ and ~~their properties have been thoroughly~~ analyzed ~~in the literature~~ [3].

6. ~~It should be borne in mind that~~ there may be a high environmental contamination level [Smrkolj, 2005].

7. Finally, ~~it is known that~~ in archaeological waxes, hydrocarbons sublimate over time and ~~it cannot be excluded that~~ a similar phenomenon may occur in a thin film of oil [Bakali, 2014].

8. ~~Another proposal is presented~~ in [12], ~~where~~ the authors use a time-series analysis with travel speed simulation to predict future trajectories. ~~In particular,~~ they use a process based on range querying with spatial-temporal constraints on moving object database.

5.4 reducing the word count: titles

Make the following titles more concise.

The development of a CAE tool for the prediction of the steady state and transient behavior of orbit annular machines (20 words)
= A CAE tool for predicting the steady state and transient behavior of orbit annular machines (15 words)

1. Design of a hydraulic system for liquid packaging.
2. An investigation into the modeling of telephony data flows.
3. A study of a novel system for solving the three-bus problem.
4. An investigation into the modeling of the coffee roasting process.
5. A study of a novel hydrogen production and energy conversion system.
6. An investigation into the long-term effects of a perennial fiber crop, ramie [Boehmeria nivea (L.) Gaud.], on the chemical characteristics and organic matter of soil.
7. The design and development of a system for biomass production and energy balance.

1. A hydraulic system for liquid packaging.
2. Modeling telephony data flows.
3. A novel system for solving the three-bus problem / A novel solution to the three-bus problem.
4. Modeling the coffee roasting process.
5. A novel hydrogen production and energy conversion system.
6. The long-term effects of a perennial fiber crop, ramie, on the chemical characteristics and organic matter of soil.
7. A system for biomass production and energy balance.

5.5 replacing several words with one preposition or adverb

Replace the part in bold with just one preposition.

1. The table shows all the information **relevant to** costs.

2. This engine is more powerful **with respect to** with the previous version.

3. This takes place **for the period of time between** May to July.

4. Unfortunately, some deficiencies remain **in connection with** the performance.

5. This can be done **by means of** a dedicated piece of equipment.

6. Several papers on this topic appeared **in the course of** the last century.

7. These are used **for the purpose of** estimating the costs.

8. There is no information **regarding** this topic in the literature.

9. Some deficiencies remain **in relation to** the mode of discovery.

10. There are serious problems **on account of the fact that** the performance is erratic.

(1) on	(6) during
(2) than	(7) for
(3) from	(8) on, about
(4) with	(9) in
(5) with / by	(10) because / since / as

5.6 replacing several words with one adverb

Replace the part in bold with just one just one adverb.

1. This can be achieved **in a satisfactory way**.
2. **From a conceptual point of view**, the task is quite difficult to visualize.
3. **In the normal course of events**, such occurrences are rare.
4. **As a consequence of this**, we were unable to ...
5. It can search for solutions **in an incremental way**.
6. Since these calculations are *in general* redundant, we decided to ...
7. **It is interesting to note that** x = y.
8. **It is probably the case that** x is higher than y.
9. **It is immediate to see that** this is a much simpler solution.
10. Installation is carried out **in an automatic fashion**.

(1) satisfactorily

(2) conceptually

(3) normally

(4) consequently

(5) incrementally

(6) generally

(7) interestingly

(8) probably / likely

(9) clearly

(10) automatically

78

5.7 replacing several words with one word

Replace the phrases in bold with one word.

1. This one was bigger **with respect to** the other one.

2. This has **made it possible for** us to do …

3. It is possible **on the condition that** the cost is specified as well.

4. We **found it useful to consider** the three cases separately.

5. The time interval between **the time instant at which** a packet arrives and **the time instant at which** it leaves.

6. **From now on** these will be referred to as X and Y.

7. This **allows us to reduce** the complexity.

8. This is an interesting theory and is similar to **the one proposed by Smith**.

9. This was done **by means of** the tracking feature.

10. Recently **the use of** such methods has led to unexpected side-effects.

11. This happened **owing to the fact that** the samples were not uniform.

12. **It should furthermore be underlined** that the samples were not uniform.

13. There are two journals explicitly devoted to **the study of** branding strategies.

14. This number **was equal to** 47% in low-income countries.

(1) than	(8) Smith's
(2) enabled	(9) using / with
(3) provided	(10) using
(4) considered	(11) since / because
(5) when, when	(12) note
(6) hereafter	(13) [studying]
(7) reduces	(14) was

5.8 replacing a *verb + noun* construction with a single verb: 1

Find a one-word equivalent for the following:

> To reach a conclusion = to conclude

1. to achieve an improvement   _____
2. to carry out a test   _____
3. to cause an increase   _____
4. to effect a reduction   _____
5. to execute a search   _____
6. to exhibit a performance   _____
7. to give an explanation   _____
8. to implement a change   _____
9. to make a comparison   _____
10. to perform an installation   _____

(1) improve
(2) test
(3) increase
(4) reduce
(5) search

(6) perform
(7) explain
(8) change
(9) compare
(10) install

5.9 identifying verb and noun clauses

Select the sentence (a or b) that contains a verb clause or adverbial clause rather than a noun clause. Note that verb and adverbial clauses are generally preferred by native speakers.

(a) Crops were selected after <u>the screening</u> of the germination energy. [NO! this is a noun phrase].

(b) Crops were selected after the germination energy <u>had been screened</u>. [YES! this is a verb phrase].

(1a) The activity was not revealed after an overnight incubation of gels.
(1b) The activity was not revealed after the gels had been incubated overnight.

(2a) However, it presents some subtle logical features that mean that it has to be identified separately.

(2b) However, it presents some subtle logical features that impose a separate identification.

(3a) The frequencies will be calculated in order to verify their probability of occurrence during an earthquake.

(3b) The frequencies will be calculated in order to verify how likely they are to occur during an earthquake.

(4a) The activity of these animals was documented from 1950 until their extinction in 1971.

(4b) The activity of these animals was documented from 1950 until they became extinct in 1971.

(5a) It is important to consider the color of the slides and the form of their presentation.

(5b) It is important to consider the color of the slides and how they are presented.

(6a) They failed to take into account the possible actions and their interleaving.

(6b) They failed to take into account the possible actions and how they might interleave.

(7a) The main feature of this notation is that concurrent constraints can be used.

(7b) The main feature of this notation is the possibility of the use of concurrent constraints.

(8a) Section 1: Choosing the Notation.

(8b) Section 1: The Choice of the Notation.

(9a) Its solution typically requires considerable computational power.

(9b) Considerable computational power is required to solve it.

(10a) Interestingly, aggregates were found, thus suggesting that the cells may be involved in the alterations.

(10b) Interestingly, aggregates were found, thus suggesting an involvement of the cells in the alterations.

(1b) The activity was not revealed after the gels *had been incubated* overnight.

(2a) However, it presents some subtle logical features that mean that it *has to be identified* separately.

(3b) The frequencies will be calculated in order to verify how likely *they are to occur* during an earthquake.

(4b) The activity of these animals was documented from 1950 until *they became extinct* in 1971.

(5b) It is important to consider the color of the slides and *how they are presented*.

(6b) They failed to take into account the possible actions and *how they might interleave*.

(7a) The main feature of this notation is that concurrent constraints *can be used*.

(8a) Section 1: *Choosing* the Notation.

(9b) Considerable computational power is required *to solve* it.

(10a) Interestingly, aggregates were found, thus suggesting that the cells *may be involved* in the alterations.

5.10 replacing a *verb + noun* construction with a single verb: 2

Replace the verb + noun *construction with a single verb.*

The patient <u>demonstrated</u> a strange <u>behavior</u>.
= The patient behaved strangely.

1. Table 1 shows a comparison between X and Y.

2. As shown in Figs. 13 and 14, the calculation makes a prediction that X will …

3. The arrival of the X occurred at …

4. The evaporation of X then takes place.

5. An increase in X of 30% was achieved.

6. Even though this results in a significant reduction in overheads.

7. The design of X is aimed at supporting multimedia services.

8. This device is used as an interface which allows the transfer of x to y to be performed.

1. Table 1 *compares* X and Y.

2. … calculation *predicts* that X will …

3. X *arrived* at …

4. X then *evaporates*.

5. X *thus increased by* 30%.

6. Even though this *significantly reduces* the overheads.

7. X *is designed to support* multimedia services.

8. This device is used as an interface *to transfer* x to y.

5.11 replacing a noun phrase with a verb or *can:* 1

Replace the phrases in bold with a verb or can. *Rearrange the sentence where necessary.*

A <u>comparison</u> of x and y *was made.*
= x and y were compared.

1. The user **has the ability to** re-send the email again with the original attachment.
2. **The configuration of the automatic pricing is done** using the xyz file.
3. This takes place **during the setup of connections** between X and Y.
4. The first step is **the choice of** X and Y.
5. This is determined by **the choice of** X.
6. This happens **at their first occurrence.**
7. This highlighted **a much better performance of X** compared to Y.
8. **After the start up of the system,** X is …
9. Particular care has been taken **in the design of X.**
10. X **presents customisability features**.

1. The user **can** re-send the email again …
2. **The automatic pricing is configured** using the xyz file.
3. This takes place **while connections** between X and Y **are being set up.**
4. The first step is **to choose** X and Y.
5. This is determined by **choosing** X.
6. This happens **when they first occur.**
7. This highlighted **that** X performed …
8. **After the system has been set up,** X is …
9. Particular care has been taken **in designing X.**
10. X **can be customized**.

5.12 replacing a noun phrase with a verb or *can*: 2

Replace the phrases in italics with a verb or can. *Rearrange the sentence where necessary.*

1. ***The installation of the system is done*** automatically.

2. ***The management of these systems can be done*** by …

3. This section ***contains an explanation of*** the various parameters.

4. These methods will be used ***for an investigation of*** the properties of …

5. These are used as markers, ***thus making possible their detection*** at low levels.

6. The aim of this document is ***the evaluation of*** new solutions for …

7. Note that ***the insertion of a new value can be performed by the user***.

8. I will outline the essential characteristics ***of application, registration and protection of*** a trademark.

9. Market liberalisation ***permits the coexistence of several firms offering the same type of service***.

10. The production of mass and standardized goods and services ***is a characteristic of*** the 19th century.

---

1. ***The system is installed*** automatically.

2. ***These systems can be managed*** by …

3. This section ***explains*** the various parameters.

4. These methods will be used ***to investigate*** the properties of …

5. These are used as markers, ***so that they can be detected*** at low levels.

6. The aim of this document is ***to evaluate*** new solutions for …

7. Note that ***the user can insert a new value***.

8. I will outline the essential characteristics ***for applying, registering and protecting*** a trademark.

9. Market liberalisation ***means that several firms offering the same type of service can coexist***.

10. The production of mass and standardized goods and services ***characterizes*** the 19th century. / ***The 19th century is characterized by the production of mass and standardized goods and services.***

5.13   replacing nouns with verbs in titles of papers

Rewrite these titles so that the words in italics are replaced with a verb.

1. The **Specification** and the **Evaluation** of Educational Software in Primary Schools.

2. Methods for the **Comparison** of Indian and British Governmental Systems in the 19th century.

3. A Natural Language for Problem **Solution** in Cross Cultural Communication.

4. Silicon Wafer Mechanical Strength **Measurement** for Surface Damage **Quantification.**

1. **Specifying** and **Evaluating** Educational Software in Primary Schools.

2. Methods for **Comparing** Indian and British Governmental Systems in the 19th century.

3. A Natural Language for **Solving** Problems in Cross Cultural Communication.

4. **Quantifying** Surface Damage by **Measuring** the Mechanical Strength of Silicon Wafers.

5.14   identifying whether link words could be deleted

*Read the two versions of the Abstract below. Which of the two versions
do you prefer? Do the words in bold in Version 2 really help you to understand? Or do they just distract the reader?*

VERSION 1

This paper presents the ECHO model, which was developed using XML.
When a system works in online mode, it communicates with the Echo
system in order to obtain the XML metadata of the audiovisual documents
from the database. Once the user has found a relevant document, the URI
of the document is obtained and passed to the metadata editor. To retrieve
the document, this URI is sent to the ECHO database, which returns the
XML document metadata. Each instance of the model is implemented by a
set of XML files, each corresponding to an instance of an entity of the
model. This means that ...

VERSION 2

This paper presents the ECHO model, which was developed using XML.
Specifically, when a system works in online mode, it communicates with
the Echo system in order to obtain the XML metadata of the audiovisual
documents from the database. **In particular**, once the user has found (by
means of the video retrieval tool) a relevant document, the URI of the
document is obtained and, **furthermore**, passed to the metadata editor.
Subsequently, to retrieve the document, this URI is sent to the ECHO
database, which returns the XML document metadata. **To be precise**,
each instance of the model is implemented by a set of XML files, **thus**
each instance corresponds to an instance of an entity of the model.
Consequently, this means that ...

The link words in Version 2 serve no real purpose. The text can be understood easily
without them.

5.15   deleting unnecessary link words: 1

Delete any unnecessary link words (highlighted in bold) in the following text.

With the reduction in price of digital devices for multimedia production, audiovisual material is progressively becoming ubiquitous. DVD, Digital TV, and the Internet are some examples of sources of audiovisual contents in digital form. ***Moreover***, nowadays anyone can buy a digital video camera and can produce material, which can then be easily distributed and published. ***Furthermore***, this process will be even more simplified with the advent of digital camcorders able to produce video already compressed in MPEG-2 or MPEG-4 format. ***On the other hand***, all this information will practically remain wasted without the means to actually find it. To make a comparison with textual content, it is as if we had all the pages of the Internet available as single images, and we were unable to apply some effective OCR tool to extract the textual information. ***However***, the situation for the audiovisual content is even worse, since OCR tools are quite effective as well as cheap. ***Instead***, information from audiovisual content is much more difficult to extract. ***In fact***, audiovisual information can be found in speech, in audio (e.g., music, explosions, etc.), in keyframes (which are images), or in moving objects in video. ***Furthermore,*** a sequence or a scene can be associated with a textual annotation (as with texts, there are titles for chapters and sections) and can contain faces of well-known people, objects, etc.

They could all be removed, The only exception is the last 'furthermore' as it indicates that the writer is adding additional information to the previous sentence. 'However' might also be useful to introduce the concept outlined in the sixth sentence.

5.16   deleting unnecessary link words: 2

Below is a revised version of the text in the previous exercise with the link words removed. Note how by introducing new paragraphs the need for a linker is even less and the message is more immediate.

With the reduction in price of digital devices for multimedia production, audiovisual material is progressively becoming ubiquitous. DVD, Digital TV, and the Internet are some examples of sources of audiovisual contents in digital form.

Nowadays anyone can buy a digital video camera and produce material, which can then be easily distributed and published. This process will be even more simplified with the advent of digital camcorders able to produce video already compressed in MPEG-2 or MPEG-4 format.

All this information will be wasted without the means to actually find it. To make a comparison with textual content, it is as if we had all the pages of the Internet available as single images, and we were unable to apply some effective OCR tool to extract the textual information.

The situation for the audiovisual content is even worse, since OCR tools are quite effective as well as cheap. Information from audiovisual content is much more difficult to extract.

Audiovisual information can be found in speech, in audio (e.g., music, explosions, etc.), in keyframes, or in moving objects in video. A sequence or a scene can be associated with a textual annotation (as with texts, there are titles for chapters and sections) and can contain faces of well-known people, objects, etc.

5.17 deleting unnecessary link words: 3

Delete any unnecessary link words (highlighted in bold) in the following text. Note: Unlike in the previous exercise where nearly all the link words were redundant, in this case most serve a purpose.

English is significantly more concise than French. **Hence**, it usually takes fewer words to explain a given concept in English. **Furthermore**, because the English language has not been "protected" – some would say fossilized – by the equivalent of the Académie Française, it is much easier to create new words to describe new concepts. This is very useful in the scientific or technical world where new concepts are constantly appearing as science progresses. A case in point is the word "redshift" in astronomy. It was invented when the expansion of the Universe was discovered in the 1920s and refers to the displacement of galaxy spectra towards longer wavelengths induced by this expansion. A correct French translation would be "déplacement (spectral) vers le rouge", quite a mouthful! **Needless to say** French scientists nearly always prefer to use the term redshift, even in front of a French audience. English speakers – and even more so Americans – love conciseness.

As an example, the word e-mail was instantaneously adopted by the worldwide academic community almost as soon as the internet was invented in the 1980s. It took several years and much debate for the Académie Française to formally introduce its French equivalent "courriel" in the Dictionnaire. **As a result,** few French people use "courriel", while e-mail has been accepted worldwide as a convenient and concise contraction of "electronic mail". The price to pay for conciseness is a certain loss of accuracy. Usually, French, with its rather strict grammatical rules is quite accurate and, when employed correctly, leaves little room for ambiguity (this is one of the reasons why French has been the favorite language of diplomats for so long as it is preferable to avoid ambiguities when drafting a peace treaty for instance!). This is not the case in English. **For instance**, because most words do not have a gender, it is sometimes difficult to decide to which noun a given adjective refers to.

All the link words could be kept with the possible exception of *hence* (at the beginning of the second sentence).

5.18 unnecessary use of *we* and *one:* 1

*Read the following pairs of sentences. Decide (1) in which cases there is a difference in meaning, (2) whether you prefer the **a** sentences (which all contain* we *or* one*) or the **b** sentences (which do not contain* we *or* one*).*

(1a) Another interesting proposal was put forward by Southern et al. [2016]. In their proposal we have a set of rules that ...

(1b) Another interesting proposal was put forward by Southern et al. [2016]. This proposal consists of a set of rules that ...

(2a) Finally, we have that $x = y$.

(2b) Finally, $x = y$.

(3a) We present a complete transform theory in Appendix 1.

(3b) A complete transform theory is given in Appendix 1.

(4a) Unlike Southern et al., we define the cost as being ...

(4b) Unlike Southern et al., the cost is defined as being ...

(5a) We may write this in the following form:

(5b) This can be written in the following form:

(6a) We close this chapter with a summary of x, y, and z.

(6b) A summary of x, y, and z follows.

(7a) One of the advantages of PCA analysis is that it enables one to classify new samples quickly.

(7b) With PCA analysis new samples can be classified quickly.

The **(a)** sentences, i.e., those sentences that use the personal pronoun (*we* or *one*) are all correct English, but if much of the paper is written in this style it can become very heavy for the reader. So where *we* refers to both the author and the reader, as in the majority of the cases in the exercise, it is better replaced with another construction. An exception is 4. In 4 the authors are making their own definition. Therefore 4b is wrong as it seems that the definition refers to what has already been established in the literature.

5.19   unnecessary use of *we* and *one:* 2

Rewrite these sentences so that all instances of we, us *and* one *are removed.*

1. First of all, we need to explain the presence of several plateaus in the graph. We can justify this behavior by bearing in mind that a given node corresponds to a set of possible clustering values. In order to clarify this concept, let us consider nodes that …

2. If we focus our attention on the nodes, we can see that they present, at the same time, a low average degree of X and a high average of Y.

3. To summarize, one can reasonably assume that X is likely to be the most important factor.

4. Observing the average values of some metrics, we can conclude that X is equal to 3.

5. In the literature one can clearly see that many of these problems have already been solved. In addition, if we focus on the comparison between the two subfigures in [36] we can easily observe that …

6. Since this latter observation may be counterintuitive, we need to underline that the presence of X does not preclude Y.

7. We suppose that the heterogeneity of these indices is mostly caused by the presence of a large number of Zs.

8. In conclusion, the structural analysis leads us to assert that X consists of:

1. First of all, the presence of several plateaus in the graph is because a given node corresponds to a set of possible clustering values. In order to clarify this concept, consider nodes that …

2. Note that the nodes simultaneously present a low average degree of X and a high average of Y.

3. To summarize, X is likely to be the most important factor.

4. The average values of some metrics show that X is equal to 3.

5. In the literature many of these problems have already been solved. In addition, a comparison between the two subfigures in [36] highlights that …

6. This latter observation may be counterintuitive, so it is worth highlighting that the presence of X does not preclude Y.

7. The heterogeneity of these indices is assumed to be mostly caused by the presence of a large number of Zs.

8. In conclusion, the structural analysis would seem to indicate that X consists of:

5.20 avoiding redundancy in introductory phrases

Make the sentences more concise by deleting the parts in italics and rear-ranging / rewriting the sentence.

1. X is an expensive item. *In this respect*, perhaps it would be better to find a less expensive substitute.

2. *As far as* the latest developments in the private sector *are concerned*, we believe that they show a marked tendency ...

3. *As far as* the top end of the market *is concerned*, there are as yet no signs of the situation improving.

4. *Regarding* Q, it was found to be unsuitable.

1. X is an expensive item, so it would be wiser to find a less expensive substitute.

2. [We believe that] the latest developments in the private sector show ...

3. At the top end of the market, there are as yet no signs ...

4. Q [,however,] was found to be unsuitable.

5.21 avoiding redundancy in references to figures, tables etc.

The following sentences are all used to describe figures and tables. Make them more concise.

1. As highlighted in the scheme reported in Figure 7, there is no change in ...

2. The mass spectrum, shown in Figure 14, proved that x = y.

3. The average amino acid composition of some animal and vegetable proteins found in art and archaeology is reported in Table 2.

4. Figure 1 shows a graphical representation of the comparison of query performances.

1. As shown in Figure 7

2. The mass spectrum (Figure 14) proved

3. Table 2 reports the ... and archaeology

4. Figure 1 shows the comparison of query performances.

5.22 rewriting unnecessarily long sentences: 1

Rewrite the following sentences so that they are more concise. Clearly there are many ways to do this, the key just shows one way (possibly the most radical).

At the beginning our research activity was mainly dedicated to the investigation of the parameters regarding ...
= Initially we investigated the parameters regarding

1. In the following diagram we show what happens when ...

2. Climatic conditions (temperature, rainfall) were also checked.

3. The objective of this document is to present ... The interested reader can find a more complete introduction in [67].

4. It is necessary to set the parameters.

5. It is possible to send comments while writing in the file.

6. In the diagram it is highlighted how the applications work.

7. Looking at the spectra presented in Figure 2, one can observe that ...

8. The procedure schematized in Figure 2 is based on a chemical treatment of the sample.

9. In the literature, there are several works that address the problem of predicting future locations [24, 36, 37].

10. The program exhibits the ability to merge data.

1. The following diagram shows what happens when ...

2. Temperature and rainfall were also checked.

3. This document presents ... See [67] for a more complete introduction.

4. The parameters must be set.

5. Comments can be sent while writing in the file.

6. The diagram highlights how the applications work.

7. The spectra presented in Figure 2 highlight that ...

8. The procedure in Figure 2 is based on a chemical treatment of the sample.

9. There are several works that address the problem of predicting future locations [24, 36, 37].

10. The program merges / can merge data.

5.23   rewriting unnecessarily long sentences: 2

Rewrite the following extracts so that they are more concise.

The research focused the comparison between the year 2011, when a severe spring frost occurred, and the 2012–2014 period, characterised by a lack of spring frosts.
= We compared 2011, when a severe spring frost occurred, with 2012–2014 when there were no spring frosts.

1. However, we should also mention that, because of the increase in the price of oil products, some public investors have started to pay attention in the last few years to renewable and non-conventional energy sources, including geothermal energy.

2. Nevertheless, in general, it can be observed that all the patients that participated in the survey achieved a significant level of improvement with regard to their motor skills.

3. In the second part, i.e., in Section IV, the coupling of gas chromatography with mass spectrometry is overviewed at the present state of the art and a comparison between our approach and preceding methods is made as well.

4. Very fundamental and valuable information has already been obtained on various chemical parameters (e.g. metals) from ice cores. Another important aspect that must be taken into account is that carbon dioxide, oxygen isotopes, and methane should be …

5. In general, the quantitative analysis of compounds that have been previously subjected to derivatisation reactions requires some remarks. The first, apparently trivial, and generally underestimated, is the necessity to use an internal standard. Then, the second is …

6. The overall effect of these three events corresponds to an increase in concentration by a factor of about 7 with respect to the concentration level observed in the period 1600–1750.

7. In the first decade of the 21st century i.e. 2001–2010, the trend in spending seems to decrease with the same slope as the previous period.

8. In these cases, analytical data may lead to draw a hypothesis about the presence of an unexpected material, and the study of selected reference materials, together with the interpretation of historical documents, allows us to confirm or not the initial hypothesis.

9. In the near future, with the increased availability of the instrumentation, it is possible that proteomics will become the preferred approach to protein determination in paintings.

10. It is fundamental to stress that most of the profiles obtained, from a quantitative point of view, do not correspond to any of the reference samples analysed, and this is in agreement with what can also be highlighted by other publications in the literature.

1. However, the increase in price of oil products has led some public investors to pay attention to renewable and non-conventional energy sources, including geothermal energy.

2. All the patients in the survey significantly improved their motor skills …

3. In Section IV we overview the state of the art in coupling gas chromatography with mass spectrometry and we compare our approach with previous methods.

4. Fundamental information has already been obtained on various chemical parameters (e.g. metals) from ice cores. In addition, carbon dioxide, oxygen isotopes, and …

5. There are three important issues related to the quantitative analysis of compounds that have been subjected to derivatisation reactions. Firstly, an internal standard must be used. Secondly, …

6. Overall these three events cause an increase in concentration by a factor of about 7 over the concentration level observed from 1600 to 1750.

7. From 2001 to 2010, the trend in spending seems to decrease with the same slope as the previous period.

8. In these cases, analytical data may indicate the presence of an unexpected material. By studying selected reference materials and interpreting historical documents, we can verify this hypothesis.

9. With the increased availability of instrumentation, proteomics may soon become the preferred approach to protein determination in paintings.

10. In agreement with the literature, most of the profiles obtained do not correspond quantitatively to any of the reference samples analysed.

5.24 rewriting unnecessarily long sentences: 3

Make the following sentences more concise.

1. In this paper we focus on the mechanisms that are available to improve performance.

2. In the next section, we give an introduction to the aspects of 802.16 that are related to QoS provisioning. The interested reader can find detailed information in ...

3. Figure 1 shows schematically that ...

4. Graphical object modeling languages have become very popular in recent years. Among those, we may cite x, y and z.

5. Several studies dealing with the evaluation and the achievement of quality in requirement documents can be found in the literature. We will briefly discuss some of them that we consider to be of particular interest to our research.

6. Smith and Jones [17] focus on the evaluation of ambiguity in natural language (NL) requirements. They start from the consideration that ambiguity in requirements is not just a linguistic-specific problem and propose the idea of checklists addressing not only linguistic ambiguity but also the ambiguity related to a particular domain.

7. Some recent market research [19] aimed at investigating the existence of a potential demand for automatic tools for requirements analysis concluded that ...

1. This paper focuses on [the] mechanisms [that are available] to ...

2. The next section overviews the aspects of 802.16 that are related to QoS provisioning – for more details see ...

3. Figure 1 shows that the ...

4. Graphical object modeling languages have become very popular including, for example, X, Y and Z.

5. Several studies on quality in requirement documents can be found in the literature, below we overview those that are particularly relevant to our research.

6. Smith and Jones [17] focus on evaluating ambiguity in NL requirements. Their premise is that ambiguity in requirements is not just a linguistic-specific problem and suggest using checklists that cover not only linguistic ambiguity but also domain-specific ambiguity.

7. Some recent market research [19] on the potential demand for automatic tools for requirements analysis concluded:

5.25 reducing length of an abstract

This abstract is nearly 300 words long. Imagine that you only have space for 100 words. Delete anything that you think is not essential information (i.e. information that might be more suitable in the Introduction).

How we speak and write generally reflects the way we think and act. This paper aims to prove that this thesis is true not only at a personal but at a national level too. To this end we analysed two languages, English and Portuguese, to verify whether the structure of the language is reflected in the lifestyle of the respective nations – in the case of English we analysed Great Britain and Australia, and for Portuguese, we analysed both Portugal and Brazil. First, we developed a methodology for establishing a Verbosity Index in the two languages. This index is based on the level of redundancy, complexity and difficulty of understanding in scientific writing. The verbosity rate for Portuguese was found to be considerably higher than for English, and was then compared to the Organisational Capacity Rate in the two countries. The latter rate is based on the difficulties due to disorganisation in typical daily bureaucratic activities (form filling, opening a bank account etc.). Our findings support the hypothesis that the simpler and more concise style and structure of the English language is evidence of an easier yet more structured society, whereas the verbose and complicated structure of Portuguese is indicative of the difficulties and disorganisation of life in both Portugal and Brazil. We also call into question the classical approach to analysing the two languages, which concludes that English is inherently more simple than other European languages. We contend that this apparent simplicity is a result of a concerted effort to make what was originally an equally complex language more readable and accessible. The results of our work should stimulate the writers of government documents around the world to express themselves more concisely and clearly. The outcome should be a considerable reduction in costs.

This paper outlines a methodology for establishing the amount of verbosity in a nation's language. The resulting Verbosity Index was then compared with the degree of disorganisation in that nation's society. Scientific writing in English and Portuguese were taken as examples, and our findings show that the simplicity of English over Portuguese, is indicative of the more organised lifestyle in Anglo countries. The assumption in the literature that English is by nature a simpler language than Portuguese, is called into question. Our results should stimulate the writers of government documents around the world to express themselves more concisely and clearly. The outcome should be a considerable reduction in costs.

5.26   reducing length of an introduction

Below are some extracts from the Introduction to a computer science paper. Do not worry, you do not need to understand the terminology. On the basis of how clear it is, choose which version, A or B, you prefer. NB the shortest version is not necessarily the best one.

	a	b
1.	In this work we propose to focus on a possible method to do …	This work focuses on a possible method to do …
2.	The challenge in Broadband Wireless Access (BWA) networks is to provide Quality of Service.	The challenge in Broadband Wireless Access (BWA) networks lies in the need to provide Quality of Service.
3.	As a result, industry and research communities are investing considerable effort in X.	Industry and research communities are thus investing considerable effort in X.
4.	The design of 802.16 is aimed at supporting multimedia services.	802.16 is designed to support multimedia services.
5.	The tool X was developed with the aim to make automatic the NL requirements analysis based on the quality model described in Section 3. The development of X has been driven by the objective to develop a tool which is modular, extensible and easily usable. For this last purpose the tool provides a user friendly graphical interface.	X was conceived to automatize NL requirements analysis, on the basis of the quality model described in Section 3. The aim was to develop a modular, extensible tool with a user-friendly graphical interface.
6.	There are several studies on quality in NL requirement documents in the literature. We overview those that are particularly relevant to our research.	Several studies dealing with the evaluation and the achievement of quality in NL requirement documents can be found in the literature. We will briefly discuss some of them that we consider to be of particular interest to our research
7.	One of the advantages of PCA analysis is that it enables one to classify new samples quickly as well as to highlight any outliers.	With PCA analysis new samples can be classified quickly and any outliers can be highlighted.
8.	The paper is structured as follows: in Section 2 a survey of the works related to the NL requirements analysis is provided.	The paper is structured as follows. Section 2 surveys the literature on NL requirements analysis.

(1) b	(5) b
(2) a	(6) a
(3) b	(7) b
(4) b	(8) b

5.27   reducing the length of the outline of the structure

The following extract is from the end of the Introduction where the authors outline the structure of the rest of the paper. It is unnecessarily long – 130 words. Reduce it to under 100 words.

The structure of the rest of this paper is organized in the following manner. In Section 1 we give a brief overview of the literature in our field. This is followed in Section 2 by a history of the English language. Section 2 is thus essential for an understanding of our contention that English is not an inherently simple language. Section 3 is devoted to a detailed description of the materials and methods that we used in our analysis of the two languages. In Section 4 our results are summarized, whereas in Section 5 (Discussion) they are discussed. Section 6 draws some conclusions based on the findings of our research. Finally, our plans for future research in this field are outlined and the paper is concluded.

Square brackets indicate further possible cuts.

[This paper is organised as follows.] Section 1 gives a brief overview of the literature. A history of the English language is presented in Section 2, which is essential for an understanding of our contention that English is not an inherently simple language. The materials and methods that we used in our analysis of the two languages are described in Section 3. We summarize our results in Section 4, and then discuss them in Section 5. Finally, after the conclusions in Section 6, we outline our plans for future research [in this field].

5.28   reducing the length of the review of the literature: 1

The sentences below come from a Review of the Literature. In each case, decide which sentence (a or b) you prefer.

(1a) X and Y are known to be characteristic of Z [Bach, 2014].

(1b) X and Y are characteristic of Z [Bach, 2014].

(2a) X and Y were once used in the Middle East and the Far East [Bakali, 2012].

(2b) The use of X and Y has been ascertained in various regions of the Middle East and the Far East [Bakali, 2012].

(3a) Also X and Y contain a small quantity of Z [Yamashata, 2013].

(3b) Also X and Y have been reported to contain a small quantity of Z [Yamashata, 2013].

(4a) In the literature the detection of X has also been reported in ceramic artifacts [Santana, 2014, McLaughlin 2015].

(4b) X has also been identified in ceramic artifacts [Santana, 2014, McLaughlin 2015].

(5a) In archaeological findings the occurrence of X may be correlated to Y [Shankar 2011, Hussein 2015].

(5b) Several authors [Shankar 2011, Hussein 2015] have suggested that in archaeological findings the occurrence of X may be correlated to Y

The most concise, without any real loss of information, are: (1) b (2) a (3) a (4) b (5) a

In each case the redundant information has been removed. The word 'literature' is not usually required if the sentence contains a reference to the literature, e.g. [Bach, 2014]. The result is that your review will be less heavy and quicker for the reader to read.

5.29   reducing the length of the review of the literature: 2

Delete any unnecessary phrases in this extract from the Review of the Literature. You only need to delete phrases – do not make any other changes.

In a very interesting paper, MacNamara (1967) stressed the need to consider the degree of bilingualism not as a unitary component, rather as a level of competence in writing, reading, speaking and listening. In this view, bilingual competence is seen as a continuum in which individuals may vary in the degree of proficiency for each of the four linguistic skills. Several descriptors have been described in the literature that are used to define proficient or less proficient bilinguals. One of the most common, as reported in many papers, describes balanced bilinguals as those who have an equal mastering of both languages (Lambert, Havelka & Gardner, 1959; Starsky and Hutch, 1970; Bobzyer Oncle, 2011). Several authors in the more recent literature have argued that balanced bilingualism is very rare (see for example the following two works: Beatens Beardsmore, 1982; Grosjean, 1997). Thus, according to the literature taken as a whole, bilingual individuals may be more dominant in one language (L1) and have their second language (L2) as the subordinate language.

In a very interesting paper, MacNamara (1967) stressed the need to consider the degree of bilingualism not as a unitary component, rather as a level of competence in writing, reading, speaking and listening. In this view, bilingual competence is seen as a continuum in which individuals may vary in the degree of proficiency for each of the four linguistic skills. Several descriptors *have been described in the literature that* are used to define proficient or less proficient bilinguals. One of the most common, *as reported in many papers*, describes balanced bilinguals as those who have an equal mastering of both languages (Lambert, Havelka & Gardner, 1959; Starsky and Hutch, 1970; Bobzyer Oncle, 1980). Several authors *in the more recent literature* have argued that balanced bilingualism is very rare (*see for example the following two works:* Beatens Beardsmore, 1982; Grosjean, 1997). Thus, *according to the literature taken as a whole*, bilingual individuals may be more dominant in one language (L1) and have their second language (L2) as the subordinate language.

5.30 reducing the length of the materials and methods

Reduce the length of this extract from the Methodology section of an engineering paper. It is currently 118 words, try to reduce it to 90 words or less.

Let us now consider a disk as shown in the schematic representation in Figure 1. The figure clearly highlights that the disk is coupled with a pin with an initial radial clearance. An approximation is introduced for the purposes of a simplification of the geometry for the analytical and numerical solutions. If one now takes into consideration Figure 2, it is immediate to see that it is now possible to decompose the vector that indicates the position of the mass. Furthermore, the centrifugal forces may feasibly be written as the sum of the two components x and y. In the following analysis for ease of discussion, the pin is assumed as being rigid and the disk as deformable.

Consider a disk as shown in Fig. 1. The disk is coupled with a pin with an initial radial clearance. An approximation is introduced in order to simplify the geometry for the analytical and numerical solutions. Figure 2 highlights how to decompose the vector that indicates the position of the mass. The centrifugal forces may be written as the sum of the two components x and y. For simplicity, hereafter the pin is assumed as being rigid and the disk as deformable.

5.31 reducing the length of the conclusions section

The following sentences come from the middle of a Conclusions section. For each sentence (1–4) decide which version A or B you prefer, on the basis of how clear and / or concise you think it is.

	VERSION A	VERSION B
1.	In this study it is concluded that compression plays an important part in ... It was found that ...	Compression plays an important part in ... In fact, it was found that ...
2.	A number of compounds are responsible for delaying the onset of ...	This work has demonstrated that a number of compounds are responsible for delaying the onset of ...
3.	The crystal structure reveals that ...	We have shown that the crystal structure reveals that ...
4.	It has been suggested in this paper that the neurons are a good marker for neuronal viability.	The neurons are a good marker for neuronal viability.

Suggested answers: (1) b (2) a (3) a (4) b

5.32   reducing the length of the acknowledgements

Cut these Acknowledgements by at least a third without removing anyone's name!

The work described in this paper is an extended version of my PhD thesis. It would not have been possible without the financial support very kindly provided by ISTI CNR, and the precious moral support of Dr. Renzo Beltrame. In addition, thanks are also due to Andrea Ceccolini, whose help was fundamental in the preparation of the Verbosity Index, the graphs and the formulas. Last but not least, I would also like to take this opportunity express a special thank you to my PhD students whose scientific writing in English convinced me of the absolute need to carry out the research whose results were presented in this work.

This work would not have been possible without the financial support of ISTI CNR, and the moral support of Dr. Renzo Beltrame. Thanks are due to Andrea Ceccolini, who helped to prepare the Verbosity Index, graphs and formulas. I also thank my PhD students whose writing in English convinced me of the need to do this research.

Section 6: Ambiguity and political correctness

TOPIC	ENGLISH FOR RESEARCH USAGE, STYLE, AND GRAMMAR	ENGLISH FOR WRITING RESEARCH PAPERS
which / who vs *that*	7.2–7.6	6.1, 6.2
-ing form	11.7–11.9	6.3, 6.4, 6.5, 15.5
a, one and *the*		6.6
pronouns + political correctness	15.7, 15.8, 15.14	6.8
use of *we* and active vs passive	10.4	7.1–7.6, 12.9
the former, the latter	15.4	6.9
and		6.12
repeating words to aid understanding	15.13	6.16

English has no gender (masculine, feminine, neuter) or case system (nominative, accusative etc.) – this means that the relations between words in a sentence may require more effort to express.

If your language has gender and / or a case system, you probably don't need to worry about your reader understanding what noun a pronoun refers to, or how the various nouns relate to each other, or which adjectives describe which nouns. When you translate into English, your sentences may become ambiguous because although <u>you</u> know what the pronouns refer to, readers may not know because they no longer have the help of gender or case. You will need to replace pronouns with their related nouns much more frequently. This means you may need to repeat the same noun many times. Don't worry about repetition – it is not a terrible taboo in English (unless you a writing literature!).

A. Wallwork, *English for Academic Research: Writing Exercises*,
DOI 10.1007/978-1-4614-4298-1_6, © Springer Science+Business Media New York 2013

6.1 repetition of words to aid reader's understanding: 1

Decide which form in bold makes the information contained in the sentence quicker and easier for the reader to absorb by not forcing the reader to have to re-read anything.

1. In [14], Gugerevic made a case for blah blah blah blah blah blah blah blah blah blah blah blah blah blah blah blah blah blah blah blah blah blah blah blah blah blah blah. Interestingly, in [15], Yang made a similar proposal to Gugerevic in which he stated that blah blah blah blah blah blah blah blah blah blah blah blah blah blah blah blah blah blah blah blah blah blah blah blah blah blah blah blah blah blah blah. The **former author's / Gugerevic's** findings thus illustrate that …

2. Sometimes prion transmission occurs through blood transfusion, contaminated surgical instruments, or skin lesions. In **the first case / a blood transfusion**, a breach of the blood–brain barrier may be responsible for the disease.

3. There are two possibilities: either $x = 1$, or $y = 1$. They are mutually exclusively and blah blah blah blah blah blah blah blah blah blah blah blah blah blah blah blah blah blah blah blah blah blah blah. **The first one / If $x = 1$ then this** implies that …

4. There are several countries involved in this project: Peru, Chile, Honduras and the Philippines, all of whom had very similar initial budgets and, in addition, were all subject to the same qualifying criteria. Note that **the latter / the Philippines** were the last to join the project which meant that …

5. Mercury is used for a variety of purposes blah blah blah blah blah blah blah blah blah blah blah blah blah blah blah blah blah blah blah blah blah blah blah blah. In the past, **this metal / mercury** was considered as being ..

In all cases the second form is the best because it prevents the reader from having to go back to the previous sentence in order to understand / remember what *former, first, latter,* and *metal* refer to. You may think such repetition is in inelegant. You should concentrate more on communicating your ideas in the simplest and clearest way possible, rather than being worried about elegance (which is more relevant in the literary world than the academic world). What this exercise also highlights is that concrete specific words are much quicker to absorb than generic and / or abstract words.

6.2 repetition of words to aid reader's understanding: 2

Decide which form (a or b) makes the information contained in the sentence quicker and easier for the reader to understand the exact meaning without having to read the sentence twice.

1. The wives were interviewed separately from the husbands as it was expected that (a) **they** / (b) **the wives** might feel intimidated by their partners.

2. The rationale for the overall sum of the various budgets is shown in Table 1 in the (a) **appendix which is taken** / (b) **appendix. Table 1 is taken** from a previous paper [Mono, 2015].

3. We rejected (a) **the samples that** / (b) **only those samples that** were contaminated.

4. Our findings are in accordance with their findings, (a) **which all show high values** / (b) **in fact both sets of findings show high values**.

5. I took my mobile phone out of my bag and then left (a) **it** / (b) **my phone** on the train.

All the (a) answers are open to possible misunderstanding, whereas the b answers are all clear.

1. *they* could also refer to *husbands*

2. *which* could also refer to *appendix*

3. use of *only those* makes it 100% clear that not all the samples were contaminated

4. *which* could also refer to *their findings*

5. *it* could also refer to *bag*

6.3 avoiding ambiguity due to use of -*ing* form: 1

Decide in which of these sentences the use of the –ing form is either po-
tentially ambiguous or simply incorrect.

1. Yoga prevents a build-up of uncomfortable physical symptoms,
   enabling you to relax more easily.

2. These impulses move from one nerve to another, ***dispatching***
   messages to the brain.

3. Doctors ***working*** in the US say that some illnesses connected with the
   heart may be cured by Biofeedback.

4. ***Watching*** TV in English, foreign students have improved their listening
   skills.

5. ***Watching*** TV for more than four hours a day can cause brain damage.

6. This is done ***clicking*** on the mouse.

7. Taxes will be lowered ***creating*** more jobs.

8. ***Reviewing*** the data, always make sure you have not left anything out.

9. I learned English ***helping*** my professor to write papers.

10. ***By eating*** vegetables alone, we cannot understand how some vegans
    do not have a deficiency in this vitamin.

1. Not ambiguous, but better: ***thus enabling.***

2. Ambiguous. Does this mean (a) the impulses move ***by dispatching messages***, or (b)
   the impulses move ***and dispatch*** messages?

3. OK.

4. Incorrect. The correct version is: ***by watching.***

5. OK.

6. Incorrect. The correct version is: ***by clicking.***

7. Ambiguous. Does this mean (a) ***by creating*** jobs (i.e. how taxes will be lowered) or (b)
   thus creating jobs (i.e. the consequence of lowering the taxes)?

8. Incorrect. The correct version is: ***when reviewing.***

9. Incorrect. The correct version is: ***by helping.***

10. Incorrect. The correct version is: ***Given that vegans only eat vegetables, we cannot ...***

6.4   avoiding ambiguity due to use of *-ing* form: 2

Rewrite the parts in italics using by, thus or when, + -ing.

1. **If we cancel** world debt, trade will increase.

2. **Every time you use** this apparatus, ensure you wear sterile gloves.

3. They stopped using sprays **and this reduced** the pollution levels.

4. The price was lowered **and consequently sales increased**.

5. We can do this **in the following manner: reinforce** Y.

1. by cancelling

2. when using

3. thus reducing

4. thus increasing sales

5. by reinforcing

6.5   disambiguating sentences: 1

Decide what is ambiguous about the parts highlighted in bold. Then, disambiguate the sentence by modifying / rewriting it. You can invent any information that you need.

1. As soon as a candidate has been reviewed by the interviewer, **he / she** shall ...

2. I have read the papers and reviewed the proposals, do you want to see **them**?

3. Thank you for your paper and the presentation **which** I have now sent to the editor.

4. To print your booking card, you will need a user name and your passport details. If you don't have **one**, then please contact ...

5. Each service is characterized by a performance parameter, as reported in Table 5, **which** describes how well the service is carried out.

6. This will not be possible, **at least in the short term**.

7. Radioactive waste is shipped in casks designed to prevent the release of the radioactive material into the environment in normal as well as accidental conditions. The way these casks are designed has been the subject of many papers. In [1] the authors state that ... Another approach is in [2], where new materials are proposed. The authors of [3] put forward an innovative idea for housing the casks in ... **The aim of the study** is to verify the scale of ...

8. On the other hand, U is the "output" quantity, either a voltage or a current, relative to a branch of immittance X_u connected between two nodes u and w, that can also be equal to ∞ and 0 **in the first and second case**, respectively.

1. Does *he / she* refer to the candidate or the interviewer? In this case, it is best to replace he / she with the appropriate noun that it refers to?

2. Does *them* refer to the papers or the proposals, or both? Again, replace *them* with the relevant noun.

3. Does *which* refer to the paper or presentation? Replace *which* with the relevant noun.

4. Does *one* refer to user name or the passport? Although it is probably obvious that *one* refers to the user name, it is still easier and quicker for the reader to understand if *one* is replaced with the relevant noun.

5. *Which* seems to refer to *table* whereas in reality it refers to *parameter.*

6. What exactly is the *short-term* – months, years, decades?

7. What *study*? Presumably the authors mean their own study. If so, then it would be better to begin a new paragraph, and to write: *The aim of our study was to ...* or *Our aim was to ...* The reader will thus be 100% clear whose work you are talking about, and it will also draw attention to your own work.

8. Without re-reading the sentence, the reader is unlikely to remember what *cases* are being referred to. It would be clearly to write: *equal to ∞ and 0 for voltages and currents, respectively.*

6.6 disambiguating sentences: 2

Disambiguate the following sentences.

1. To take our children to the party we all used our cars and then we left them there.

2. This should help to prevent piracy of CDs by Americans.

3. We investigated lions and tigers and elephants that are on the endangered species list.

4. We surveyed various immigrants: Tunisians, Moroccans and Senegalese who had entered the country before 2015.

5. After opening the program, the email can be sent to multiple recipients.

6. The fire broke out in the forest bordering the river but was extinguished before any major damage could be done by the local fire service.

7. Instructions for use: These articles are poisonous. If there are children in the house, keep them locked up safely.

8. A full range of games were presented for the men with no balls (e.g. darts, diving, bungee jumping).

9. Being over 500 years old, Dr. Alvarez handled the painting with great care.

10. Preserved in a frozen state, Professor Chang examined the samples.

1. To take our children to the party we all used our cars and then we left the **children** there.

2. This should help to prevent piracy of **CDs produced by American artists**.

3. We investigated lions and **tigers, and also those elephants** that are on the endangered species list. / We investigated lions, tigers, and elephants. **All three types of animals** are on the endangered species list.

4. We surveyed various immigrants – Tunisians, Moroccans and Senegalese – **all of whom** [*assuming that we are talking about all three nationalities*] had entered the country before 2015. / We surveyed various immigrants who had all entered the country before 2015: Tunisians, Moroccans and Senegalese.

5. **After the program has been opened**, the email can be sent to multiple recipients. / **When you have opened the program**, you can send the email to multiple recipients.

6. The fire broke out in the forest bordering the river but was extinguished **by the local fire service** before any major damage could be done.

7. These substances are poisonous. If there are children in the house, keep the **substances** locked up safely.

8. A full range of games **requiring no balls** were presented for the men (e.g. darts, diving, bungee jumping).

9. The **painting** was over 500 years old, so Dr. Alvarez handled it with great care.

10. Professor Chang examined **the samples, which** were preserved in a frozen state.

6.7 pronouns and political correctness

Decide which of the following pronouns should be inserted into the spaces in these three extracts from a paper: (a) he, (b) she, (c) he / she, (d) they, (e) his, (f) her, (g) his / her, (h) their, (i) one, (j) Ø

If a scientist feels it necessary, therefore, to publish in English in order to reach this worldwide audience, does this mean that the scientist whose native language is not English is at a disadvantage in attempting to get **(1)** ___ work published and accepted? Certainly, there does seem to be evidence that scientists from developing countries do find it more difficult to get **(2)** ___ work published than those in developed countries.

Clearly many articles are rejected because of inadequate command of English and nobody would expect a journal to publish an article which was full of grammatical mistakes in English. However, does this mean that a foreign scientist, not a native speaker of English but **(3)** ___ who reads and writes English fluently, is at a disadvantage in getting published? Given the number of articles published by those whose native language is not English in prestigious international journals, it is obvious that being a native speaker is not an absolute prerequisite to publication. If a foreign scientist then has a good command of English, does it matter that **(4)** ___ publishes in English rather than French, Finnish or Chinese? .... This comparison of the very different languages of English and Chinese, which traditionally have been said to have quite distinct rhetorical styles, did not find great differences between the languages. Regardless of language, all papers followed the classic pattern of such introductions, as put forward by John Swales in **(5)** ___ widely accepted CARS (Creating a Research Space) model.

1. **Ø** or **their** – it is now considered by many to be perfectly acceptable to use **their** to refer to a generic person even if the noun it refers to is singular (*a scientist*). **his / her** would also be possible, but is unnecessary. In reality, the author of the paper (a man) used the feminine pronoun **her**. This choice of using the female pronoun to refer to a generic person is quite rare, and is usually found amongst more politically aware authors.

2. **Ø** or **their** – in this case no one can argue with the use of **their** as the noun it refers back to is plural (*scientists*). **his / her** would also be possible, but is unnecessary.

3. **One** or **Ø.**

4. **They** (the same reasoning as in 1 applies here). The author of the paper used **she**.

5. **His** (simply because Swales is a man).

6.8 non-use of masculine terms for generic situations: 1

Rewrite so that the sentences are gender neutral (i.e. no use of he / his / him *to refer to both males and females).*

> ***The user*** can then drag ***his*** files into a folder.
> = ***Users*** can drag ***their*** files into a folder.
> = ***The user*** can drag ***his / her*** files into a folder.

1. Using the same techniques as we have used, the reader can compute the other 25 products using his calculator.

2. If a student sees the paper of another student, she may feel that ...

3. Since ancient times, man has depicted his life through objects painted onto a variety of formats.

4. Those interviewed who did not get along with their neighbors claimed that their neighbor was often rude and that he would occasionally resort to violence.

5. When asked about their doctor, most interviewees said he was a friend of the family. Interestingly, with regard to any nurses that interviewees said they had met in hospitals, the majority said that she often then became a friend.

6. Anyone doing a presentation must send his presentation in PDF format.

7. Man's origins are still not fully understood.

8. Many Englishmen drink too much alcohol with consequent liver and kidney complaints in later life.

9. Each student must have his own laptop.

1. ... using **a** calculator / using **his / her** calculator / using **their / a** calculator.

2. ... another student, **he / she** may feel that ...

3. Since ancient times, **humans have depicted their life / we have depicted our lives** through objects ...

4. ... and that **he / she** would occasionally resort to violence.

5. ... most interviewees said **he / she** was a friend of the family. Interestingly, with regard to any nurses ... the majority said that **he / she** often then became a friend.

6. ... must send **their** presentation / must send **his / her** presentation in PDF format.

7. **Our** origins / The origins of the **human species** / The origins of **human kind** ...

8. Many **English people** drink too much alcohol ...

9. Each student must have **their** own laptop / have **his / her** own laptop.

6.9   non-use of masculine terms for generic situations: 2

Write a 10 word summary of the following text. Note: This text was written in 1906, well before the age of political correctness. Make sure you only use he *in reference to the male sex.*

Every individual is what he is through his interaction with his surroundings, physical and social. As a social being he enters into all sorts of relations with other human beings, and with the world in general, and as he grows older the range and scope of these relations increase. Cut away the social side of a person and the individual is reduced to a social nothing. His development and success in life depend upon the fullness and wisdom with which he enters into suitable relations with the world in which he finds himself. To do this implies that he understands these relations and can interpret them generously and liberally.

6.10   non-use of masculine terms for generic situations: 3

Write a 10 word summary of the following text. Note: This text was written in 1906, well before the age of political correctness. Make sure you only use he *in reference to the male sex.*

The teacher is like a guide, and the pupil like a traveller in an unknown country. The traveller knows where he wants to go, but knows neither the way nor the exact character of the place he wishes to reach. The guide knows both, and plans the journey so as to set out from where the traveller now is and to reach where he desires to be, and that by the best way. Such plotting of the journey is analogous to the teacher's laying down his course of instruction in any subject with its order of topics and arrangement of matter. But unless the traveller – that is the pupil – take the journey himself, nothing is accomplished. Many a lesson is too much like a guide describing the journey to the would-be traveller, who sits and listens but does not leave his chair to undertake it. In other words the guide himself laboriously takes the journey again and again, but the traveller that should be remains inert. In short, no matter how admirably a lesson is planned, there is no real methodical teaching unless the pupils by their own efforts pass along the road traced for them; for, as has been said, true teaching is nothing but arousing and directing the learning activity of another.

Section 7: Paraphrasing and avoiding plagiarism

TOPIC	ENGLISH FOR RESEARCH USAGE, STYLE, AND GRAMMAR	ENGLISH FOR WRITING RESEARCH PAPERS
plagiarism		10.1
copying generic phrases		10.2
quoting by paraphrasing		10.4, 10.5, 10.6
referring to other authors		14.3, 16.11
personal (we) vs impersonal style	10.1–10.4, 15.8	7.2–7.6, 12.9
avoiding repetition from the Abstract in the Introduction		13.6
avoiding repetition in the Conclusions		18.3, 18.4

Paraphrasing is a very important skill as it will help you to avoid

- Plagiarism (cutting and pasting from other papers without acknowledging the source i.e. taking the credit for work that others have done).
- Word-for-word repetition of what you wrote earlier in the paper.

Many of the skills of paraphrasing are practised in other sections of this book:

- Being concise, avoiding redundancy.
- Changing nouns to verbs.
- Changing from a personal form to an impersonal (or vice versa).
- Changing the word order.

When paraphrasing, concentrate on the concept rather than the actual words used by the original author.

A. Wallwork, *English for Academic Research: Writing Exercises*,
DOI 10.1007/978-1-4614-4298-1_7, © Springer Science+Business Media New York 2013

118

7.1 deciding what is acceptable to cut and paste

Which of the words and phrases in bold would you be able to cut and paste into your own work without stating the source because they are sufficiently generic that they could apply to any situation?

In this research project I will consider bilinguals as (1) *"those who use more than one language or a dialect in their everyday life"* (Grosjean, 2010). The inclusion of dialects is particularly relevant here, (2) *as part of the project involved* Italian participants. (3) *In Italy, different dialects are spoken in different regions.* These dialects are not just mild inflections from the mother tongue, but proper languages that may significantly differ in syntactic, semantic and phonological properties. For example, (4) *someone from Sicily who speaks Sicilian and Italian should be considered as bilingual as someone from Barcelona who speaks Catalan and Spanish*. As in most of the Italian regions a dialect can be spoken for historical and cultural reasons, (5) *we may say that* a considerable proportion of Italians, especially in older generations, are bilinguals.

(6) *What is bilingualism?* I asked this question to an artist, the one who painted the work represented at the beginning of this chapter. She replied: (7) *"Bilingualism is my fourth dimension. It is the way I see things without boundaries, without communication constraints. Bilingualism is a space in which culture flies freely and the mind expands to new fascinating territories."*

Perhaps this definition of bilingualism is too romantic. However, I feel that (8) *it captures the very nature of being bilingual* in modern times. According to Beatens Beardsmore (1982) the term bilingualism has an "open-ended semantics". (9) *No definition can really explain* the complexity of the cognitive, social, educational and cultural factors that are embedded in those who embarked on a bilingual life. In this first chapter (10) *I will attempt to describe what is* bilingualism in the contemporary world, (11) *how it is studied*, and (12) *why it is important to understand* crucial cognitive mechanisms that support it in the human brain.

You should be OK to cut and paste the following: 2, 3, 5, 6, 9, 10, 11, 12

7.2   quoting statistics

Which of the statistics and findings in bold would you be able to use in your paper without making a reference to the literature or attributing a source?

The growing interest in bilingual or multilingual speakers is not surprising if we think that more than half of the world's population – (1) *about 3.5 billion people* – regularly speak (2) *more than one language* (Grosjean, 1982, 2010). As far as Europe is concerned, the European Commission recently published a report (2006) in which a large sample of European citizens were asked how many languages they spoke other than their mother tongue. (3) *Fifty-six percent of the people in 25 countries* replied that they could have a conversation in a second language, and 28% replied they spoke a third. (4) *Great Britain is one of the most "monolingual" countries in Europe*; nonetheless, (5) *38% of those polled* replied they could speak a second language.

These figures, though impressive, do not tell us much about a potentially bilingual or multilingual population that appears to be a tremendously heterogeneous group. Were the languages learnt early in childhood or later? Are the additional languages used in everyday life? How competent are these people in their second language? These (6) *three basic questions* are themselves enough to transmit even to the naïve eye how difficult studying bilinguals could be. As Grosjean (1998) pointed out "… working with bilinguals is a more challenging enterprise [than studying monolinguals].

Only the first (1), would be acceptable to quote without any source, as the world's population is a statistic that is in the public domain.

120

7.3 paraphrasing by changing the parts of speech

Write definitions of the following branches of human biology and medicine shown below. Note: do not use exactly the same grammatical structure or the same verb or noun in more than one definition.

Example: Anatomy is the study of how the body is structured and the way in which the various components are linked together.
= An anatomist studies the structure of the body and how its parts fit together.
Note the changes made in terms of grammar and / or use of synonyms:
- *Anatomy* (the science) > *anatomist* (scientist).
- *The study* (noun) > *studies* (verb).
- *Is structured* (verb) > *the structure* (noun).
- *Components* (noun) > *parts* (noun).
- *Fit* (active verb) > *are linked* (passive verb).

SCIENCE	AREA OF STUDY
1. biochemistry	chemical processes occurring in and around body cells
2. cardiology	problems relating to heart and blood vessels
3. genetics	DNA, chromosomes and genes + inheritance mechanisms
4. neurology	disorders of nervous system
5. psychiatry	mental illness – prevention and treatment

7.4 paraphrasing by changing nouns into verbs

Replace the phrases in italics with a verb and make any other necessary changes.

The presence of mixtures of saccharide materials make *the identification of a plant gum in a paint sample a difficult task.*
= The presence of mixtures of saccharide materials make *it difficult to identify a plant gum in a paint sample.*

1. *The use of* a microscope is essential for *a full comprehension of* the technique.

2. In certain environments this could *lead to an enhancement in* the lipid preservation.

3. The anaerobic bacteria can *cause a strong degradation of* the wood.

4. The amount formed *is strictly dependent* on the degree of oxidation, thus the values observed *present a high variability* and are influenced by many factors.

5. Samples were directly monitored *for the observation of* the morphological characteristics.

6. *The assessment of this index was carried out* by means of the correlation function.

7. The heating of the probe can be carried out in two different ways:

8. The main drawbacks are the increase in volume and weight of the residue *which causes the loss of the advantage of the incineration process*, and the *production* of a material that might still be very hazardous for the environment.

9. This solution implies *the reaching of* a consensus among these processes.

10. The authors wish to thank the Department of Political Sciences for *the setting up and coordination of* the project.

122

1. [*Using*] a microscope is essential for *fully understanding / in order to fully understand* the technique.

2. In certain environments this could *enhance* the lipid preservation.

3. The anaerobic bacteria can *considerably degrade* the wood.

4. The amount formed *strictly depends* on the degree of oxidation, thus the values observed *are very variable* and are influenced by many factors.

5. Samples were directly monitored *[in order] to observe* the morphological characteristics.

6. *This index was assessed* by means of the correlation function.

7. *The probe can be heated* in two different ways:

8. The main drawbacks are the increase in volume and weight of the residue which *means the advantage of the incineration process is lost*, and *a material is produced* that might still be very hazardous for the environment.

9. This solution implies *reaching* a consensus among these processes.

10. The authors wish to thank the Department of Political Sciences for *setting up and coordinating* the project.

7.5   paraphrasing by changing the parts of speech and word order: 1

Rewrite the sentences below so that (1) the word order is different (where possible), and (2) at least one element changes from, for example, verb to noun, or noun to verb, or active to passive, or adjective to adverb.

The examples below show <u>two</u> different ways to change the sentence. However, you only need to find <u>one</u> way. No key is given for this exercise as there are many ways to paraphrase each sentence
X is different from Y in a number of respects.
= There are a number of important differences between X and Y.
= X differs from Y in several fundamental ways.
This tool is targeted at end-users.
= The target of this tool is end users.
= End users are the target of this tool.
This survey provides a summary of the relevant literature.
= This article aims to do widen current knowledge of this topic.
= Summarizing the most pertinent papers in the field is the focus of this article.

1. Our experiments confirm previous results [Wiley 2009].

2. We found much higher values with respect to those reported by Pandey [2000].

3. These discrepancies are negligible due to the fact that …

4. To the best of our knowledge no other authors have found that $x=y$.

5. In conclusion, our work demonstrates that $x=y$.

6. Figure 1 clearly shows that these values reach a peak when $x=y$.

7. There is a possibility that dissimilar evaluations would have arisen if the focus had been on x instead of y.

8. Many attempts have been made [Kim 2009, Li 2010, Hai 2011] aimed at improving performance.

9. As far as we know this is the first time that this system …

7.6   paraphrasing by changing the parts of speech and word order: 2

Follow the instructions to the previous exercise.

1. The samples were prepared as required by current norms.

2. A great deal of attention must be paid when handling the samples.

3. This method suffers from a number of pitfalls.

4. We would like to thank the following people for their support, without whose help this work would never have been possible:

5. It is very likely that participants may have answered the questions incorrectly.

6. The reasons for this result are not yet entirely understood.

7. Despite the limitations of this method, and consequently the poor results in Test 2, our findings do nevertheless suggest that …

8. Given that this was only a preliminary attempt, it is hardly surprising that there were some discrepancies.

9. The benefits in terms of performance far outweigh the additional costs.

10. As anticipated, there were some discrepancies due to the massive amount of data being processed.

7.7 finding synonyms: verbs 1

Write at least one synonym for each of the words / phrases in bold.

1. Last century this procedure was ***considered to be*** the most ...

2. Previous work has ***only focused on*** addressing the symptoms rather than the cause.

3. Concerns have ***arisen*** which question the validity of ...

4. This paper ***outlines*** a new approach to ...

5. The aim of our work was to ***further*** current knowledge of ...

6. Vitous [2015] has ***provided*** a new definition, in which ...

7. A growing body of literature has ***examined*** [Ref].

8. An increase in the number of cases was first ***noted*** by ...

9. Experiments with this system were ***conducted*** in 2009 by a group of researchers from ...

10. He ***claims*** that ...

1. viewed as / seen as
2. been limited to / failed to address
3. been raised
4. proposes / describes / presents
5. extend / widen / broaden
6. put forward / proposed
7. investigated / studied / analyzed / evaluated
8. reported / found
9. carried out / performed
10. argues / maintains / suggests / points out / underlines

7.8 finding synonyms: verbs 2

Write at least one synonym for each of the words / phrases in bold.

1. Many experts now **contend** that rather than using Pappov's approach it might be more useful to …

2. To **assess** whether plastic could be converted into gold we …

3. The set up we used **can be found** in [Ref 2].

4. Our experimental set up **bears a close resemblance to** the one proposed by Smith [2014].

5. The apparatus **consists of** three main parts.

6. The interface can easily be **customized** to suit all requirements.

7. The resulting ad hoc device **can** function in various fields.

8. Having this system **enabled us** to incorporate several new parameters.

9. The interviewees were **divided** into two groups.

10. No significant difference was **revealed** between the two methods.

1. believe / argue

2. determine / check / verify / determine

3. is reported / is detailed

4. is very similar to / is reminiscent of / is based on

5. is made up of / is composed of

6. tailored / adapted / modified

7. is able to / has the capacity to

8. allowed us to / permitted us to / meant that we could

9. split / broken down

10. found / identified / detected / observed / highlighted

7.9   finding synonyms: verbs 3

Write at least one synonym for each of the words / phrases in bold.

1. We **began** this project three years ago.
2. This **underlines** just how important this system is.
3. This **confirms** previous findings in the literature …
4. Further tests carried out with this system **confirmed** our initial findings.
5. As **expected,** our experiments prove that.
6. The cost of this system **could account for the fact that** it is rarely used.
7. This research has **raised** the need for further investigation.
8. As was **mentioned** in the Methods, …
9. Table 2 **proves** that this system is **...**
10. Figure 1 **presents** the data on the new system.

1. started / initiated / commenced
2. highlights / stresses / proves / demonstrates
3. supports / lends support to / substantiates
4. corroborated / concurred with
5. anticipated / predicted / forecast / hypothesized

6. be the reason for / explain why
7. given rise to
8. stated / noted / discussed / reported
9. shows / demonstrates / illustrates / highlights
10. reports / shows / details

7.10   finding synonyms: nouns 1

Write at least one synonym for each of the words in bold.

1. This has many **uses** in the field of ...

2. A major **defect** of this procedure is ...

3. In this **report** we ...

4. A recent review of the literature on this **topic** [2012] found that ...

5. Southern's group [5] calls into question some past **assumptions** about this procedure.

6. The method is essentially the same as that used by Kirk [2009] with some **changes**.

7. This component is fully compliant with international **norms**.

8. The software **application** used to analyze the data was SoftGather (Softsift plc, London).

9. The main **criteria** for selecting the samples was not mentioned at all.

10. In all cases **patients'** consent was obtained.

1. roles / applications

2. difficulty / drawback / disadvantage / flaw

3. paper / review / study

4. subject / matter / area

5. hypotheses / theories

6. modifications / alterations / adjustments

7. regulations / standards

8. program / package

9. reasons / rationale

10. subjects' / participants'

7.11 finding synonyms: nouns 2

Write at least one synonym for each of the words in bold.

1. In the first **step** of the process …

2. This new equipment has the **ability** to outperform all previous versions.

3. Our procedure is a clear **improvement** on current methods.

4. Our study provides **additional support for** alternative methods for treating this disease.

5. Apart from this slight **discordance,** the result is confirmation of …

6. There are several possible explanations for this **result.**

7. The results point to the **likelihood** that the species will be extinct within 5 years.

8. Further experimental **investigations** are needed to estimate **…**

9. These findings suggest the following **directions** for future research:

10. An important **issue** to resolve for future studies is **…**

1. phase / stage
2. capacity / potential
3. advance
4. further evidence for / considerable insight into
5. discrepancy / disagreement / non-alignment
6. finding / outcome
7. probability
8. tests / studies
9. opportunities
10. matter / question / problem

7.12 finding synonyms: adjectives

Write at least one synonym for each of the words / phrases in bold.

1. It is **straightforward** to verify that ...

2. Malaria is the **main** cause of ...

3. Greening the Internet has become a **central** issue in ...

4. Many hypotheses regarding this system appear to be **ill-defined.**

5. In their **seminal** paper of 2001, Peters and Jones ...

6. Kamos's [23] assumptions seem to be **realistic.**

7. Their approach is not **well suited to** ...

8. The **traditional** approach to sample collection is to ...

9. Our results were **disappointing.** However, ...

10. One **possible** application of our technique would be ...

1. easy / trivial

2. leading / primary / major

3. important / critical

4. unfounded / not well grounded / unsupported / questionable / disputable / debatable

5. groundbreaking / cutting edge

6. well-founded / well-grounded / plausible / reasonable / acceptable

7. appropriate for / suitable for

8. classical / normal / usual

9. unsatisfactory / below expectations

10. potential / promising

7.13 finding synonyms: adverbs and prepositions 1

Write at least one synonym for each of the words / phrases in bold.

1. **Since** the focus of the study was on a new system, we decided to …

2. There has been some disagreement **concerning** whether x is equal to y or not.

3. Several authors have attempted to define emotional intelligence, but **as yet** there is still no accepted definition.

4. The fonts**, i.e.** the form of the characters, are of various types.

5. **Little** is known about **…**

6. Many experts contend, **however**, that this evidence is not conclusive.

7. Statistical significance was analyzed **by using** SoftGather.

8. The aim of this system is to increase performance. **Consequently** we.

9. We chose this particular apparatus **because** it is inexpensive.

10. The samples were prepared **as described by** Jude [2010].

1. Given that / As
2. regarding / with regard to
3. currently / at the time of writing
4. that is to say
5. Not much / Very little
6. instead / on the other hand /

7. through the use of / via
8. As a result / Therefore / Thus / Hence
9. on account of the fact that / since
10. in accordance with / according to / following / in line with

7.14 finding synonyms: adverbs and prepositions 2

Write at least one synonym for each of the words / phrases in bold.

1. ***Almost*** two-thirds of the participants (64%) commented that ...

2. In response to Question 1, ***most*** of those surveyed indicated that ...

3. ***Interestingly,*** only 7–8 year olds were able to find the answer.

4. There were no significant differences between the two systems ***in terms of*** cost.

5. ***Overall,*** our results show this machine outperforms all the others on the market.

6. The correlation between the two procedures is ***worth noting*** because ...

7. ***In contrast to*** earlier findings [Castenas, 2009], we ...

8. ***Despite*** the lack of agreement, we believe our findings compare well with ...

9. ***Although*** there was some inconsistency ...

10. We hope that our research will be helpful in solving this difficulty. ***At the same time*** we believe that ....

---

1. Just under / Approximately.

2. Nearly all / The majority.

3. Surprisingly / Unexpectedly.

4. With regard to Z / As far as Z is concerned.

5. Taken as a whole / Generally speaking / With a few exceptions.

6. Interesting / Of interest / Noteworthy / Worth mentioning.

7. In contradiction with / Unlike.

8. Notwithstanding.

9. Even though / Despite the fact that.

10. In addition / Further / Furthermore.

7.15   paraphrasing by changing word order

Rewrite these sentences by putting the part in bold at the beginning of the sentence. Make any changes that you feel are necessary.

1. There are several categories of race and ethnicity. **These include Hispanic, American Indian and Filipino.**

2. There are three categories of **rendering techniques.** These are A, B, and C.

3. Someone who spends their day thinking about existential problems is called **a philosopher.**

4. Someone who spends their day thinking about **existential problems** is called a philosopher.

5. The usual length of **the rod** is two meters.

6. The usual length of the rod is **two meters.**

7. The disease may be caused by **water pollution, contaminants in food etc.**

8. The categorisation combines **the ideas from previous taxonomies.**

9. It is still not fully understood **how the brain works**.

10. A courier delivered **the package**.

1. **Hispanic, American Indian and Filipino** are examples of the various categories of race and ethnicity.

2. **Rendering techniques** can be subdivided into three categories: A, B, and C.

3. **A philosopher** is someone who spends their day thinking about existential problems.

4. **Existential problems** are what a philosopher spends his / her day thinking about.

5. **The rod** is usually two meters long / in length.

6. **Two meters** is the usual length of the rod.

7. **Water pollution, contaminants in food and such like** may cause the disease.

8. **The ideas from previous taxonomies** are combined in the categorisation.

9. **How the brain works** is still not fully understood.

10. **The package** was delivered by courier.

7.16 replacing *we* with the passive form

Imagine you have written the Materials and Methods below. You then discover that your chosen journal does not allow the use of personal pronouns. Where possible and appropriate, rewrite the parts in bold by removing all instances of we *and* our.

(1) **In the first part of our study, we analysed** the length of 50 European Union documents written in English and Spanish, to confirm whether documents which purport to be exact translations of the same subject, vary substantially in length. The difference was not significant: Spanish documents were typically in the region of 5% longer.

Not convinced by this result, (2) **we then decided** to do a more detailed study. One thousand scientific articles written in English and the same number written in Spanish were scanned using a conventional high resolution scanner. A 'Word Parser' was then used to analyse the articles in terms of: word length, sentence length, frequency of use of nouns rather than verbs, the use of impersonal phrases and passive form, and the frequency of particular punctuation marks.

The two languages were compared on the basis of the number of occurrences of these elements. For example, (3) **we assumed that** there would be a direct correlation between the length of words and sentences and the reader's understanding of such sentences, i.e. the shorter the sentence (the quicker and deeper the understanding). In addition, (4) **in a previous paper [2012] we had hypothesized** that the use of verbs rather than nouns, and personal rather than impersonal constructions, leads to more fluent, concise and comprehensible sentences. This information was then used to establish the difficulty in understanding the authors' (i.e. the authors of the papers being analysed) individual concepts and overall meaning.

(5) **To test our hypothesis,** (6) **we gave a selection of the sample articles** to a panel of 10 professional referees and proof-readers (all native English speakers). The sample articles were all written in English, but either by native speakers or Spaniards. To check whether non native speakers might actually find the more complex language easier to understand (as it might reflect the conventions and style of their own native tongue), (7) **we assembled a panel of 10 non-native referees**.

(8) **We also asked both panels** to note down the time (T) it took them to read a particular paragraph. They then rated their understanding (U) in a range of zero to three (0=nil understanding, 10=total understanding), and also the amount of energy (stress S) they believed to have consumed in coming to such an understanding.

(9) **With these data, we were then able** to formulate a Verbosity Index (VI):

$$kV(s) = \frac{q_1 S(s) + \sigma_2 T(s)}{\beta_3 U(s)}$$

where *V* is the verbosity of a sentence s. In fact, *V* is directly proportional to its stress *S* and to its time *T*, while it is inversely proportional to its level of understanding *U*.

(10) **We believe that our verbosity index can be used** to test the level of comprehensibility of any government document from any country.

1. In the first part of the study, an analysis was made of the length …

2. It was then decided.

3. It was assumed that.

4. In a previous paper [2012] we had hypothesized – this should not be changed as replacing *we* with, for instance, *the authors*, might confuse readers as they might think you are talking about other authors.

5. To test our hypothesis – *our* could be changed to *the*, but again the reader might not be sure whose hypothesis you are talking about.

6. A selection of the sample articles was given to.

7. A panel of 10 non-native referees was assembled.

8. Both panels were asked.

9. With these data, it was then possible.

10. We believe that our verbosity index can be used – this could be changed to *The verbosity index could be used* however the original version is less strong (*we believe* is an example of hedging, you are suggesting that this is just your opinion and others might feel differently).

7.17   making a summary: 1

Write a 20–40 word summary of the following text by B Alvarez. Decide what parts you might need to quote directly, i.e. which words might be better expressed by the original author rather than being paraphrased by you.

Researchers often believe they do not need to consider style when writing scientific papers. They consider style a matter of 'decorating' their prose to make it more attractive to the reader. In our survey of 1,000 papers written by Ph.D. students we found that this is not the case. It is much more fundamental than that and involves such things as point of view and sentence structure.

We tested six different approaches to writing papers. We found that the best was to instruct writers as follows: As you begin to organise your thoughts and your findings, decide who your reader is. It's even useful to imagine a particular person, in some cases a colleague, in others a student. What attitude should you assume? Are you trying to instruct and explain or to inform and persuade? Specialists in the same field will be familiar with your subject and its particular jargon and so won't need a lot of "prompting." You should ask yourself what essential information or meaning you want to put across in your paper.

We believe that writers should overcome the idea that they must adopt a certain 'official' style when writing technical papers. There is no one correct or 'official' voice. It is a fallacy that serious scientific journals do not accept papers written in the first person.

Many researchers aspire to an objectivity they believe is obtained by using impersonal constructions or the passive voice. They want to lay emphasis on the experiment or results and not on the observer. Scientific findings are no more or less valid because they are expressed by an identifiable author. Our survey of the literature of early scientific work highlighted that some of the greatest men of science, such as Einstein, Darwin and Louis Pasteur, were also gifted writers, unafraid to report their findings in the first person. Even today their papers exude a degree of warmth and immediacy.

7.18   making a summary: 2

Write a 20–40 word summary of the following text written by B Alvarez. Decide what parts you might need to quote directly, i.e. which words might be better expressed by Alvarez rather than being paraphrased by you.

While watching a film we unconsciously make hundreds of logical connections that enable us to follow the story line easily. We certainly don't think about the hours of film which have been cut out. Readers too make connections as they move from sentence to sentence, paragraph to paragraph.

When papers reflect a clear, logical progression of ideas, the reader follows the argument without excessive promptings like: *It is worthwhile noting that …, As a matter of fact …, Experience teaches us that …* Likewise, the overuse of such words and expressions as *hence, therefore, thus, it follows that, in fact, indeed,* and *namely* can be tiresome.

We found that many students fill their papers with redundant expressions because they think it makes their writing sound more impressive, or sometimes because they simply do not have much to say.

Why do so many writers use long, empty phrases instead of short clear ones? Perhaps it is because as George Orwell suggests "it is easy." He goes on to explain that it is easier – even quicker, once you have the habit – to say *It is not an unjustifiable assumption that* than to say *I think …*

A key finding in our work was that particularly when students are writing in a hurry they fall into a pretentious, Latinised style. Tags such as *a consideration which we should do well to bear in mind* or *a conclusion to which all of us would readily assent* will save many a sentence from coming down with a bump. By using stale metaphors, similes and idioms, it was revealed that students could save themselves much mental effort. But this at the cost of leaving their meaning vague, not only for their readers but for themselves.

7.19   making a summary: 3

Write a 10–20 word summary of the following text, which comes from a freely available NASA document on writing technical reports.

You will be given, say, 20 minutes in which to present to visiting scientists a review of your research. Your first reaction is to insist that you need more than 20 minutes. Your arguments will be rejected, however, so you eventually proceed complainingly to your task. With great ingenuity you apply yourself to outwit your stubborn and arbitrary boss by getting an hour's material into a 20-minute talk. You concoct long and brilliantly comprehensive sentences for your discussion, your organize all of your numerical results so that you can present them in rapid-fire order, and you lay out 15 slides, each crammed with detailed information. Unfortunately, when you first read through your prepared speech, it takes nearly 30 minutes instead of 20. You are not very disturbed, however, because you are confident that, with a little practice, you will be able to increase your speaking rate until the delivery time is down to the required 20 minutes.

All that is missing from your approach is consideration for the central figure – the man in the audience. After 2 minutes of your talk he will be rapidly developing mental indigestion; after five minutes he will have lost the thread of your discussion; and during the remainder of your talk he will simply concentrate on hating you or on trying to sleep.

7.20   making a summary: 4

Write a 10 word summary of the following text. Note: This text was written in 1906 before the age of political correctness. Make sure you only use he *in reference to the male sex.*

In its crudest form education consisted in training the child in the pursuits – hunting, fishing, fighting, etc. – necessary to enable him to maintain himself and his family when he should reach adult life. As occupations became more specialized the training took the general form of teaching the boy the craft or trade of his father and the girl the household duties performed by the mother.

But as communities became more organised the conception of the child as the future citizen became dominant, and, as a consequence, the idea that education is intended to train loyal and useful citizens overshadowed the conception that it is a means of benefitting the individual.

Section 8: Defining, comparing, evaluating and highlighting

TOPIC	ENGLISH FOR WRITING RESEARCH PAPERS
differentiating your work from other authors' work	Chapter 7
comparing your work with other authors' work	17.3, 17.4, 17.8, 17.11
highlighting your findings	Chapter 8, 16.6, 16.8, 16.11, 17.9,

A. Wallwork, *English for Academic Research: Writing Exercises*,
DOI 10.1007/978-1-4614-4298-1_8, © Springer Science+Business Media New York 2013

8.1 writing definitions 1

The definitions in the table have been mixed up. Rewrite them including, where necessary, an appropriate article (the, a / an) at the beginning of the sentence and incorporating 'is a' after the class.

	CONCEPT	CLASS	RELATIVE PRONOUN OR EQUIVALENT
1	*oxygen*	person	who spends a lot of their time looking for funds for projects.
2	university	tool	*which is essential for our survival*.
3	researcher	country	which has revolutionized the world.
4	USA	*gas*	which has made vast quantities of money through cutting-edge technologies.
5	Internet	company	where it is believed by many that everyone has the same opportunities.
6	Apple	place	that is widely considered to be man's best friend.
7	gold	animal	where in theory people go to study but in reality often spend most of their time just having fun.
8	dog	metal	that tends to go up in value during financial crises.

1. Oxygen *is a* gas which is essential for our survival.

2. A university is a place where in theory people go to study but in reality often spend most of their time just having fun.

3. A researcher is a person who / that spends a lot of their time looking for funds for projects.

4. The USA is a country where it is believed by many that everyone has the same opportunities.

5. The Internet is a tool which / that has revolutionized the world.

6. Apple is a company which / that has made vast quantities of money through cutting-edge technologies.

7. Gold is a metal which / that tends to go up in value during financial crises.

8. A dog is an animal which / that is widely considered to be man's best friend.

8.2   writing definitions 2

Using a maximum of 30 words, write your own definitions of for five of the following. Include articles (the, a / an) *where necessary.*

A **good teacher** is passionate about their subject and manages to pass on this passion to their students through clear explanations that the students can easily understand and relate to.

1. good teacher

2. communism

3. equilateral triangle

4. India

5. microscope

6. polygamy

7. plastic

8. sheep

9. true happiness

10. vitamin B

8.3   writing definitions 3

Extend the definitions you wrote in the previous exercise by using one of the following techniques. For each definition use a different technique.

(a) Make the definition more specific.

(b) Give a concrete example.

(c) Describe the components and / or basic operating principles.

(d) Explain the origin of the word.

(e) Outline the purpose or usage of the concept / object defined.

(f) Compare and contrast the concept / object with a similar concept / object.

8.4 making generalizations

Expand on the generalizations listed below.
Useful link words: in fact, this is because, consequently, in addition.

The **Internet** is one of the most important inventions in the last 100 years. In fact, it has completely changed the way we live and has given everyone access to vast amounts of information that would have been inconceivable a few decades again.

1. Stereotypes are often misleading.

2. Water scarcity will be the biggest cause of war in the next few decades.

3. Social, academic and business networks are becoming increasingly important.

4. The gap between the rich and the poor shows no signs of getting narrower.

5. Most manual work will one day be done by robots.

6. History repeats itself.

7. Politicians are essentially only interested in lining their own pockets.

8.5 confirming other authors' evidence

The Evidence column below represents the way a certain topic is perceived by some, though not the majority, of researchers. The main proponents of this line of thinking are in the Author / s column. You have found evidence (right-hand column) that supports the Authors. Write two or three sentences comparing your findings with the Authors' findings. Highlight the importance of your findings, inventing any additional information you need.

	AUTHOR / S	EVIDENCE	YOUR FINDING SUPPORTING CLAIM
1	Smith et al. [2012]	politicians tell the truth	Lie detectors used with politicians from 17 countries indicated that in 87% of cases, the truth is told
2	Chang & Li [2013]	First telephone invented in China in 1845 (not by Bell in 1876)	Discovery in the vaults of two Beijing museums of prototype telephones all dated between 1845 and 1851
3	Carmen [2012]	Vaccinations could be given by being mixed in milk, and then drunk, rather than through injections	34 patients given milk vaccinations. Control group given injected vaccinations. No difference in outcome
4	Guyot [1969]	Americans never landed on the moon	Investigation into engines of Apollo rockets indicates that they could only have flown for a max of 3 km

1. Contrary to other results in the literature, Smith et al. found that politicians do in fact tell the truth [2012]. We corroborated their results by using lie detectors with politicians from 17 countries, which indicated that in 87% of cases, the truth is told. This has important implications when voters make decisions as to which party they wish to see elected.

2. Although there is strong evidence that Bell was the first person to invent the telephone, Chang and Li [2013] claim that it was invented in China thirty one years earlier. Our discovery in the vaults of two Beijing museums of prototype telephones all dated between 1845 and 1851, confirms Chang and Li's claim. Our findings lend further credence to the fact that many inventions that have been assumed as being invented in the West, already existed the East decades or centuries before.

3. In 2012, Carmen published an innovative study on the use of vaccinations administered in milk. We repeated their experiments with a group of 34 children given milk vaccinations. A control group of 34 children was given injected vaccinations. Our age group was older than Carmen's: 8–10 year olds, rather than 4–6 year olds. No difference in outcome was found either between our results and Carmen's, or between the milk- and injection-administered vaccinations. The findings of these two studies make a convincing case for using milk, which would also represent a less expensive solution for national health services, and would certainly be appreciated by both children and their parents alike.

4. Immediately after the Americans had landed on the moon in 1969, the French scientist, Guyot, published evidence that it would have been impossible to put a man on the moon and that the photographs taken of the astronauts were in fact made in a Hollywood studio. Our simulation of the functioning of the engines of the Apollo rockets proves conclusively that they could never have reached the moon, in fact the rockets would only have been able to fly for 3 km. This finding corroborates other research highlighting the fear that the USA had of losing out in the 'technological war' with the USSR.

8.6   stating how a finding is important

Instead of telling your readers that a finding is interesting or significant, you should show them how it is interesting or significant. Show readers what they need to know to come to their own conclusion about the result. Based on the information given in the first column of the table below, rewrite the sentence in the second column so that this information becomes more explicit and would thus stand out from other information given in the Results / Discussion section. Invent more information if you wish. Do number 1 first (Solid A), then compare with the key (which contains just one possible answer, yours may be quite different). Then copy the style / approach of the key to do numbers 2–4.

YOUR FINDINGS	WHAT YOU SAID ABOUT YOUR FINDINGS
1 When cold, Solid A was hard, brittle and difficult to shape. When heated to 1,200 °C Solid A became soft and malleable, and changed color. No one has previously thought of heating Solid A.	Interestingly, the characteristics of Solid A when in a cold state changed radically when heated (see Table 1).
2 Solar panels take 15 years to become cost effective, not 10 years as estimated by Sun and Chang [2013]. Producing solar panels consumes ten times more energy and materials than the energy needed to extract petroleum from the sea bed to produce the same amount of power.	It was found that solar panels take 15 years to become effective, rather than the previous estimation of 10 years by Sun and Chang [2013]. In terms of parity of power, the production of solar panels requires considerably more energy than extracting petroleum.
3 Artifacts recently found at the British Museum reveal that the British Isles were colonized by the Chinese in 8,000 BPE (before present era). Comparison of modern day Celtic (language spoken in Britain before English) and Chinese reveals substantial similarities in syntax.	Using artifacts from the British Museum, we revealed that the British Isles were previously a Chinese colony. This is supported by a comparison of the Chinese and Celtic languages.
4 Playing with Barbie dolls is intellectually more productive than playing with Lego. Brainscans on over 10,000 boys and girls aged between 5 and 8 revealed that brain activity was higher in the girls when playing with Barbie, than in boys playing with Lego. Implications: boys should be encouraged to play with girls' toys.	Brainscans on over 10,000 boys and girls aged between 5 and 8 revealed that brain activity was higher in the girls when playing with Barbie, than in boys playing with Lego.

1. We believe that this is the first time that Solid A has been heated. The results of our heating tests indicate that when Solid A is heated to 1200 °C it loses its hardness and brittleness, and becomes malleable. This has very important implications: previously Solid A had not been considered for applications where the ability to shape the solid is crucial (such as antiseismic materials for skyscrapers). In addition, we found that the color changes, which may be relevant in the design and production of household equipment using Solid A.

2. Our findings highlight that unlike what was previously thought, solar panels take an additional five years before becoming cost effective (15 years rather than 10 years). Perhaps even more importantly, the cost in terms of global warming of producing solar panels as opposed to extracting petroleum is higher – not lower – than previously thought.

3. We present some outstanding new evidence that Britain was once a Chinese colony. Artifacts (including Chinese pottery from the Second Pre-Dynasty) found only last year in the archives of the British Museum, reveal that the Chinese settled in the British Isles some 10,000 years ago. We confirmed this incredible finding by making a comparison of texts from modern day Celtic languages with the first written examples of Chinese. These comparisons revealed that the two languages share 67.54% of syntactic structures.

4. Playing with Barbie dolls is intellectually more productive than playing with Lego. Counterintuitively, brainscans on over 10,000 boys and girls aged between 5 and 8 revealed that brain activity was higher in the girls when playing with Barbie, than in boys playing with Lego. It would thus seem, counter to previous research [Syco 2014] that the brain effort in terms of imagination in playing with dolls is greater than the effort required in a combination of the imagination and motor skills needed to construct with Lego. Our conclusion is likely to cause an upset in the research world: boys should be encouraged to play with girls' toys.

8.7 highlighting why your method, findings, results etc. are important

Expand the four sentences in bold, so that the reader understands why something is 'significant', 'interesting', 'innovative', 'remarkable' etc. Invent whatever you want.

This method would certainly represent a significant step forward ... *as it would enable the use of much smaller sample sizes and would thus be around 40% less invasive than current procedures.*

1. Our model is very innovative ...

2. These results are very interesting ...

3. This increase in performance is remarkable ...

4. Our findings have wide implications ...

1. Our model is very innovative ...... due to / in terms of / concerning the approach that was used, which as far as we know, has not been applied before. In fact, it can be used to estimate ...

2. These results are very interesting ... ... since they highlight that ... . because they represent an increase of 20% in yield compared to ... ... as they could pave the way to several applications.

3. This increase in performance is remarkable ... in fact, the x index is three times higher than the y index. ... Using this method helps to do x. In addition / Also / Further / Furthermore, it indicates that ... it leads to a large / considerable / substantial change in the ...

4. We believe that our methodology has many applications ... ... in the pharmaceutical field as / since / in fact / because they allow x to be produced in a single step. ... In fact our method could help the scientific community by ... These findings have wide implications, particularly in the field of ...

8.8   highlighting your findings

When you have a finding / conclusion that is counterintuitive and / or goes against what has so far been established in the literature, you need to highlight your finding so that readers can immediately (1) see the finding on the page, (2) understand its importance. The Current Thinking column below represents the way a certain topic is currently perceived in the litera- ture, and the main proponents of this line of thinking are in the Author / s column. Your controversial finding is in the last column. Write two or three sentences contrasting your finding with previous knowledge and also high- lighting your finding's importance. Clearly, you will need to invent informa- tion to support your findings and their level of significance.

Example answer for 1:
For several milleniums it has been widely held that one plus one is equal two. This, we believe, has been caused by a series of miscalculations due to the deceptively easy nature of the calculation. Our finding that one plus one is in fact equal to three, thus goes against the current. However, it is based on making the calculation using Generation 5 Computers, which clearly reveal that …

	AUTHOR / S	CURRENT THINKING	YOUR CONTROVERSIAL FINDNG
1	thousands of authors	$1 + 1 = 2$	$1 + 1 = 3$
2	Hhot, Kold	Global warming increasing.	Global warming now slowing down.
3	Princip, Gomez	First World War triggered by assassination of Archduke Franz Ferdinand in Sarajevo	First World War triggered by US interests in slowing down growth of European economy
4	thousands of authors	tomato is a fruit	tomato is a vegetable
5	thousands of authors	cigarettes are bad for the health	cigarettes are only bad for certain body organs. Other organs, such as the brain and ears, benefit from tobacco intake
6	Glee	the World Wide Web is the work of Tim Berners Lee	the WWW was designed by North Korean housewife, Romi Choson
7	Markoni & Knockia	cell phone use in cars has increased number of accidents due to distraction	younger generation much more capable of multi- tasking, use of cell phone has no effect on accident rate

8.9   comparing the literature

Complete sentences (a) and (b) so that they mean the same as the preceding sentence.

   Example: Many people believe in the existence of a greater force.

   (a) A greater force is believed to exist by many people.
   (b) It is believed by many people that a greater force exists.

1. The value of these products has been hypothesized as being much higher than previously thought.

   (a) It has been hypothesized …

   (b) Values previously estimated for these products …

2. Not all scientists consider that genetic engineering is a good idea.

   (a) Genetic engineering is …

   (b) There is no general consensus …

3. Some researchers have questioned the fact that global warming is a man-made phenomenon.

   (a) Global warming as a man-made phenomenon …

   (b) It has been questioned …

4. According to some research, homeopathic medicine only has a placebo effect and has no real medical value.

   (a) Homeopathic medicine has been found …

   (b) It has been suggested that …

5. Smith and Jones determined that destiny has a considerable effect on the life of most individuals.

   (a) Destiny has been determined to …

   (b) The life of most individuals is …

6. Irradiation of food products will increase by 180% over the next five years according to Choe et al.

   (a) Choe et al. have proposed that …

   (b) It has been …

7. Our ability to learn languages may be hereditary (Lingo and Huang, 2013).

   (a) It is thought that …

   (b) Our ability to learn languages is thought …

8. Vente et al. (2022) have revealed that wind power is ten times more efficient than previous estimates.

   (a) Wind power has been …

   (b) According to Vente et al. …

9. Dalek (2018) expects the appearance of UFOs to increase rapidly in the near future.

   (a) The appearance of UFOs is …

   (b) A rapid increase in the appearance of UFOs has been …

Note: square brackets indicate that the phrase contained therein is optional and / or could be located in more than one place in the sentence. There may be other possible answers not given in this key.

(1a) It has been hypothesized that the value of these products is much higher than previously thought.

(1b) Values previously estimated for these products may have been much lower than their true value.

(2a) Genetic engineering is not considered by all scientists to be / as being a good idea.

(2b) There is no general consensus on whether [or not] genetic engineering is a good idea.

(3a) Global warming as a man-made phenomenon has been questioned [by some researchers].

(3b) It has been questioned whether [or not] global warming is a man-made phenomenon.

(4a) Homeopathic medicine has been found to have only a placebo effect and to have no real medical value.

(4b) It has been suggested that homeopathic medicine only has a placebo effect and has no real medical value.

(5a) Destiny has been determined [by Smith and Jones] to have a considerable effect on the life of most individuals [(Smith and Jones)].

(5b) The life of most individuals is affected considerably by destiny according to Smith and Jones.

(6a) Choe et al. have proposed that irradiation of food products will increase by 180% over the next 5 years.

(6b) It has been proposed that irradiation of food products will increase by 180% over the next five years [Choe et al.].

(7a) It is thought that our ability to learn languages may be hereditary (Lingo and Huang, 2013).

(7b) Our ability to learn languages is thought to be [possibly] hereditary (Lingo and Huang, 2013).

(8a) Wind power has been revealed to be ten times more efficient than previous estimates.

(8b) According to Vente et al. (2022) wind power is ten times more efficient than previous estimates.

(9a) The appearance of UFOs is expected to increase rapidly in the near future (Dalek, 2018).

(9b) A rapid increase in the appearance of UFOs has been predicted / forecast by Dalek (2018).

8.10   comparing contrasting views

Choose three scenarios from the table below. Write a short paragraph, highlighting the two opposing views. Also, add your own particular perspective for each case by stating to what extent (and why) you support the claim or counterclaim.

Example answer for 1:
Geldov et al. claim that Industrialized nations need to send more aid to Africa. A completely opposing view was expressed by Njimi, who states that African countries need to sort out their own problems – financial aid only makes things worse. We believe that it is possible to take the best of both approaches by using aid in more subtle ways that do not fall into the trap of making the recipient country dependent on aid.

	AUTHOR / ORGANIZATION	CLAIM	AUTHOR / ORGANIZATION	COUNTER-CLAIM
1	R. Geldov et al.	Industrialized nations need to send more aid to Africa	Njimi	African countries need to sort out their own problems – financial aid only makes things worse.
2	Schwarz	study of 60 cats in Germany; conclusion: cat is man's best friend	Santana et al.	study of 82 dogs in Mexico: dog is man's best friend
3	Berlusconi, Mafiosovic	voters will ignore corruption and immorality in politicians provided tax rate kept low	Sensato	low tax rate no guarantee that voters will ignore criminality in the government
4	Wordsworth	English is best equipped to be the language of science	Zapata	Esperanto much better equipped
5	Ferrari, Lotus	speed limits on motorways and autobahns in Europe actually increase the number of accidents	Zhang	analogous studies in China reveal that optimum speed limit should be 180 km
6	Igno and Ramus	education spending should be cut on "non-productive" areas of research such as philosophy	Conphusio and Zokrate	philosophy has been the keystone to civilization in the east and west for thousands of years
7	many doctors	there is absolutely no proof that homeopathy works from a medical point of view	Deek and Hehd	homeopathy works for both humans and animals
8	Harris, Goldemein	peer review by experts in the field is the best method of judging suitability of papers for publication	Siedelsen	peer review by third-year Ph.D. students who are not experts in this specific field (but are experts in a related field) is the best method

8.11 comparing your methodology with other authors' methodologies

Explain how your methodology differs from existing methodologies regarding typical academic activities. Justify why your methodology is superior. Add / Invent any details.

Example answer for 1:
Many experts on how to write research papers state that it is best to write the paper in chronological order. In our approach, the writer begins with what he / she finds to be the easiest section, which is often the Materials and Methods. We believe that our approach works well because the writer is immediately given a boost in confidence, which enables him / her to then face the difficulties in writing the more complex parts (e.g. the Abstract and the Discussion).

	ACADEMIC ACTIVITY	YOUR METHODOLOGY	DIFFERENCE FROM EXISTING METHODOLOGIES
1	Writing a research paper	Begin with easiest section (generally Methods)	Write all the paper in chronological order
2	Preparing a presentation	Begin by writing a script (everything that you want to be say during the presentation), create slides on the basis of the script	Begin by creating slides, then decide what to say
3	Attending lectures	Write notes, record lecture (with lecturer's permission), concentrate at all times	Send text messages, go on FB, complete other tasks on laptop
4	Preparing for an examination	Takes notes throughout the course. Highlight important parts of notes. Revise highlighted parts	Begin revision the night before the examination
5	Responding to very critical referees	Begin by saying something positive, agree with as much of the referee's report as possible, never criticize or say anything negative	Begin by undermining referee's credibility, justify why referee's recommendations have been ignored, adopt an angry tone

8.12 comparing data in a table

The table summarizes the efficiency of two methods for learning English. Make a comparison between the two methods. In your description make sure you use:

- *At least one irregular comparative form (e.g.* worst, fewest*).*
- *All the following words:* less, fewer, more, much, many.
- *At least two adverbs (e.g.* efficiently, quickly, fluently*).*

Below is an example in which the first few rows of the table are compared.

In both methods the length of the study was the same (i.e. four months), however the number of participants in Method B was higher than in Method A: 421 and 375, respectively. Not as many words were learned in Method B than in …

	METHOD A	METHOD B
No. participants in study	375	421
Length of study	4 months	4 months
No. words learned that can be used actively	500	456
No. words that can be understood	3,000	1,500
No. of tenses learned	5	8
Level of fluency achieved	mid	low
% errors made when speaking	35%	15%
Writing ability	low	good
% understood while listening to authentic radio news	10%	20%

8.13   questioning current thinking

*The following premises and conclusions come from Abstracts and Intro-
ductions. Imagine that you are writing a paper on a similar topic. Write
short paragraphs which present a different perspective on the premise and
/ or the conclusion.*

PREMISE: Industrial nations became rich partly through the use of child
labor in their own countries during the Industrial Revolution.
CONCLUSION: Developing nations should be free to exploit child labor in
order to gain a similar degree of wealth.
PREMISE ACCEPTED, CONCLUSION QUESTIONED: While it is certainly
true that child labor helped industrial nations to produce considerable
wealth during the Industrial Revolution, this does not justify its continued
use in the 21st century. Human rights laws have advanced considerably in
the last 200 years, and child labor is simply no longer acceptable.

	PREMISE	CONCLUSION
1	China is a massive economic power with the highest population in the world.	Chinese will one day replace English as being the world's international language.
2	Industrial nations have polluted the world for two centuries.	Less stringent regulations regarding carbon emissions should be applied for developing countries.
3	Humans are more important than animals.	Animals should be used for testing products that might be harmful for humans.
4	English is the international language of science.	All articles submitted to international journals should be written in perfect English.
5	Everyone has the right to buy and consume what they wish.	It is perfectly acceptable that massive gas-consuming cars originally intended for off-the-road use should also be used on normal roads.
6	Hospitals are overcrowded and limited funds are available for health services.	Certain people should be refused urgent medical intervention if they have them-selves have caused their own health problem (e.g. through smoking, over-eating, under-eating, drugs).
7	We are all born with the same rights.	Everyone should have the right to free education at all levels and free books.
8	Top actors and top football players have skills that no others possess.	They should be paid massive amounts of money.

8.14   evaluating solutions

*Write a short paragraph to evaluate the success of the solutions
(in the table below) in accordance with their aims.*

AIM: cut carbon emissions
SOLUTION: ban all imports of fruit and vegetables
Reducing carbon emissions is clearly a good aim, as is encouraging the consumption of fruit and vegetables that are produced locally. However, if all imports of fruit and vegetables were banned this would have two major effects. Firstly, some farmers in poorer countries would see their source of income completely eliminated. Secondly, those countries that for climatic reasons only produce a limited number of fruits and vegetables, would find themselves on a limited (and unhealthy) diet of, for example, solely apples and potatoes.

	AIM	SOLUTION
1	get a good job	do a Master's or Ph.D.
2	redress balance between developed and developing countries	cut all the debts of developing countries
3	reduce consumption of alcohol and tobacco	raise taxes on these goods until a point when the price becomes prohibitive
4	reduce drug-related crimes and other drug-related issues	legalize drugs
5	stop top scientists leaving developing countries to work in developed countries	force developed countries to fund the same research in the developing country and give the scientist a similar salary as the one that he / she would have obtained in the developed country
6	make people use public transport or become part of car-pool schemes	impose a heavy fine on anyone driving a car with no passengers
7	reduce obesity	ban all fast food restaurants and the ubiquitous use of fructose as an sweetener in foodstuffs
8	improve education standards	increase the number of teachers and pay them ten times more their current salary

Section 9: Anticipating possible objections, indicating level of certainty, discussing limitations, hedging, future work

TOPIC	ENGLISH FOR WRITING RESEARCH PAPERS
anticipating possible objections, convincing readers of your point of view	8.11 9.7, 17.8
toning down (i.e. making certain words sound softer)	9.3–9.5
indicating level of certainty / probability	9.6
limitations of your work	9.9, 14.6, 17.12, 17.13
limitations of the literature	9.11, 9.12, 14.6, 17.11
hedging	chapter 9
over hedging	9.13
future work	18.6

A. Wallwork, *English for Academic Research: Writing Exercises*,
DOI 10.1007/978-1-4614-4298-1_9, © Springer Science+Business Media New York 2013

158

9.1 anticipating objections and alternative views

In your Discussion you often have to support your view against another view already contained in the literature, or against an objection that you think the referee, editor or reader might raise. Choose two or three cases in the table below. Write a paragraph in which you describe the significance of your claim / view and why you reached it. Then mention the alternative view or objection and minimize its importance. You will need to invent information.

Example for 1
Our results clearly show that mirages are the result of a temporary malfunction of the brain. This goes against the views of other authors who state that mirages are seen when light traveling through the air follows a curved, rather than a straight, line. Although the latter explanation may seem plausible, it does not explain why mirages are seen on moonless night, when there is in fact no light source. Also, the instrumentation used to assert that light travels in a curved fashion was invented in the 1950s. More modern technology [Knerd, 2012] reveals that in fact light is lazy and rarely travels at all.

	YOUR VIEW / CLAIM	ALTERNATIVE VIEW / POSSIBLE OBJECTION
1	Mirages are the result of a temporary malfunction of the brain	Mirages are seen when light traveling through the air follows a curved, rather than a straight, line
2	Intuition is superior to observation, as stated by the followers of Pythagoras	Intuition is based on previous experience of similar situations. The conclusions drawn from such previous experience may be wrong. Moral judgements are non scientific, and dangerous
3	Dinosaurs became extinct because they were tele-transported to another planet by aliens and used as weapons against an enemy nation	Dinosaurs became extinct due to a giant meteorite which caused dust that impeded sunlight reaching the earth
4	Sigmund Freud never went to university and never studied medicine. No records of attendance at university have ever been found. Dream theory was invented by his wife	Freud was the pioneer of psychoanalysis, he studied medicine at the University of Vienna, Austria
4	Lemons were originally native to Scotland and then spread to the Mediterranean. Lemon fossils have been found in many mountainous regions in the north of Scotland	*Citrus lemon* is native to Asia. Lemons were first grown in Europe in the time of the ancient Romans. It was commonly found in the Arab world and Mediterranean countries around 900 years ago

9.2 indicating level of certainty 1

Associate a percentage to the expressions below on the basis of the level of certainty that they give. Clearly, your answers can only be approximate.

It is highly likely that … 100% **90%** 75% 50% 25%

PHRASE		% LEVEL OF CERTAINTY				
1.	It is highly likely that …	100	90	75	50	25
2.	It is far from being certain that …	100	90	75	50	25
3.	It seems unlikely that …	100	90	75	50	25
4.	This may have happened because …	100	90	75	50	25
5.	This is probably due to …	100	90	75	50	25
6.	There is no doubt that …	100	90	75	50	25
7.	We believe that …	100	90	75	50	25
8.	As far as we aware …	100	90	75	50	25
9.	This cannot be the case …	100	90	75	50	25
10.	There is no reason to believe that …	100	90	75	50	25
11.	This is perhaps caused by …	100	90	75	50	25
12.	This is possibly the result of …	100	90	75	50	25
13.	This should help to …	100	90	75	50	25
14.	This must be the reason for …	100	90	75	50	25
15.	This could be due to …	100	90	75	50	25
16.	This is thought to be the consequence of …	100	90	75	50	25
17.	There is a slight possibility that …	100	90	75	50	25
18.	This is known to be the case for …	100	90	75	50	25

1. 90
2. 25
3. 25
4. 50
5. 75
6. 90
7. 90
8. 90
9. 100
10. 100
11. 50
12. 50
13. 75
14. 100
15. 50
16. 75
17. 25
18. 100

9.3 indicating level of certainty 2

Look at the theories in the first column. Indicate how certain (100–25%) you are that these theories may in fact reflect reality. Then choose three or four of the theories. Explain why you do or do not agree with them. Incorporate some of the phrases from the previous exercise.

THEORY	100%	75%	50%	25%
1 Economic crises are invariably caused by greedy and irresponsible bankers				
2 The petroleum industry has suppressed the development of water-powered cars				
3 Editors frequently reject manuscripts in order to allow another work on a similar topic to be published first				
4 Hell exists as a real physical place				
5 There is a meaning to life				
6 Airport restrictions on what passengers are permitted to take on planes were not implemented in the 2000s on the basis of a real terrorist threat but only on the perception by the public of such a threat – thus they reassured the public that something concrete was being done to protect them				
7 A cure for cancer will be found in the next 20 years				
8 People will soon be living in underwater communities				
9 It is difficult to fool a lie detector				
10 It is dangerous to swim immediately after eating				

9.4 reducing level of certainty

Make the following sentences sound less arrogant / strong.

Our approach lends itself perfectly for use by civil engineers. = *We **believe** our approach will lend itself for use by civil engineers. / Our approach **could possibly** lend itself for use by civil engineers.*

1. These results demonstrate the fundamental nature of ...

2. This factor is responsible for the increase in x.

3. Few papers have been published on x.

4. Little is known about this.

5. This is the first time that such results have been reported in the literature.

6. Our finding that x = y represents a groundbreaking discovery.

1. These results would appear to demonstrate the fundamental nature of ... / These results suggest the fundamental nature of ... / We believe that these results demonstrate ...

2. This factor is likely *[US English]* responsible for the increase in x. / This factor is likely to be *[GB English]* responsible for ... / This factor is probably responsible for ...

3. As far as we are aware / To the best of our knowledge, few papers ...

4. As far as we are aware / To the best of our knowledge, not much is known about this.

5. We believe that / As far as we are aware / To the best of our knowledge this is the first time that such results have been reported in the literature.

6. We believe that our finding that x = y is a groundbreaking discovery.

Note: *As far as we know* and *To the best of our knowledge* should only be used when making comparisons with the state of the art in the literature.

9.5   discussing the limitations of the current state of the art

Below is an example talking about limitations and filling the gap. Use this example as a model, in order to explain (a) the main limitations in the current state of the art in your field of research, (b) how your work overcomes these limitations.

The main limitation of the current approach to teaching languages is that it focuses on a syllabus that was created in the 1950s. Preliminary changes to the syllabus proposed by Keats et al. (2007) **gave promising results**. They suggested that various exercise types could be abandoned as they were not time effective in terms of what was actually learned. **Unfortunately**, no model has yet been designed to help understand the advantages and disadvantages of their idea. **Moreover**, they offered no real solutions to the problem that they had uncovered. This paper **attempts to fill this gap** by developing an analytical model of a way to teach languages that reduces the traditional five years of schooling down to just one year.

9.6   qualifying what you say

In which sentences below does the author qualify his / her results, i.e. admit that they are limited in some way?

1. Although the results are limited to X, and do not take into account Y, this can be justified from the fact that …

2. Even though our results do not show that $x = y$, they nevertheless highlight that …

3. The results are good, albeit only from a small sample.

4. Our results allowed us to calculate the value of the parameters investigated.

5. Despite the fact that our model was not able to predict the values with 100% accuracy, we believe that our achievement of 87% is a major advance on the current state of the art.

All apart from 4

9.7   dealing with limitations in your own results: 1

Read this extract from a Discussion. Note how the author reveals and deals with limitations of his results. Then do 9.8.

Although limited to just one parameter, the test results in Table 2 and the experimental points in Fig. 3 confirm the accuracy of the forecasts. For that particular parameter, the test data and the analytical results show that optimum performance is already achieved with only three workers involved. The experimental data in Table 3 *exceed the model forecasts* by approximately 40%, 23% and 17% for $n=4$, 5, and 6 workers, respectively. *These may seem quite large errors, but they can be explained by ... In fact,* given a larger workforce (which was unfeasible due to our low budget), *we would have achieved* far greater accuracy.

9.8   dealing with limitations in your own results: 2

Now underline the phrases in the Discussion below that indicate where the author wishes to show that her study (on the occurrence of tuberculosis amongst the homeless in London) may have limitations. Then do 9.9.

Our results show a high prevalence of tuberculosis (17.2 per 1000 screened) among men over 50. This is likely to be an underestimate as the screening was voluntary and a number of clients declined the screening altogether. It is well documented that homeless people face many barriers in accessing adequate healthcare services [Peters, 2011]. In addition health care may not have been viewed as a major priority – in fact, the availability of luncheon vouchers probably motivated many to volunteer for the screening carried out at our institute.

Five per cent of those interviewed admitted to tuberculosis in the past. This is significant as the risk of reactivation may have been high due to alcohol abuse, poor nutrition or hostel living conditions, as reported in [Smith, 2014]. In our study, no cases of active tuberculosis were detected among the white ethnic population under 40 or among women, although the total number of women screened (280, 14%) was relatively small.

The prevalence of tuberculosis that was found among homeless refugees was six per 1000 screened. A combination of factors such as poverty, poor living conditions (e.g. in hostels and B&Bs) and stress may have been important in explaining the epidemiology of the disease among this population. The findings of this study in relation to refugees are inconclusive and highlight the need for further research.

Our results show a high prevalence of tuberculosis (17.2 per 1000 screened) among men over 50. ***This is likely to be an underestimate*** as the screening was voluntary and a number of clients declined the screening altogether. It is well documented that homeless people face many barriers in accessing adequate healthcare services [Peters, 2011]. In addition health care ***may not have been viewed*** as a major priority – in fact, the availability of luncheon vouchers ***probably motivated*** many to volunteer for the screening carried out at our institute.

Five per cent of those interviewed admitted to tuberculosis in the past. This is significant as the risk of reactivation ***may have been [this is not a limitation, but just an explanation]*** high due to alcohol abuse, poor nutrition or hostel living conditions, as reported in [Smith, 2014]. In our study, no cases of active tuberculosis were detected among the white ethnic population under 40 or among women, ***although the total number of women screened (280, 14%) was relatively small.***

The prevalence of tuberculosis that was found among homeless refugees was six per 1000 screened. A combination of factors such as poverty, poor living conditions (e.g. in hostels and B&Bs) and stress ***may have been important in explaining*** the epidemiology of the disease among this population.
The findings of this study in relation to refugees ***are inconclusive and highlight the need for further research.***

9.9   dealing with limitations in your own results: 3

Choose two or three cases from the table below. Write a paragraph in which you: (a) describe your findings, (b) mention the limitations, (c) explain how these could be dealt with.

Example for 1
We found that a ratio of 5:1 in praise / criticism in the classroom significantly improves student performance. One limitation of our work was that studies were only made in five schools, and those were all in the same town. We thus plan to extend our studies to other schools in other towns, and possibly in other countries.

	YOUR FINDINGS	POSSIBLE LIMITATIONS	SOLUTION TO LIMITATIONS
1	A ratio of 5:1 in praise / criticism in the classroom significantly improves student performance	Studies only done in 5 schools all in the same town	Extend studies to other schools in other towns and countries
2	High-powered magnets can be used to extract metals from rock	Cost of magnets far higher than value of metal extracted	Design lower-cost magnets
3	Medicinal properties found by Druids still relevant today and show 89% efficacy	Patients given Druid medicines may have been sympathetic to use of non-traditional (alternative) medicines	Pre-interviews should be conducted to ascertain patients' views of alternative medicine
4	Benign dictatorship is the best form of government. GDP of countries with benign dictators 65% higher than those with democracies	Definition of 'benign' How does dictator take control? Who decides when his / her term is over?	Introduce benign dictators into five western democracies for a 5-year period. Compare results with previous 5-year period
5	Internet use has rewired and enhanced the human brain. People now able to speed-read effectively and quickly find information desired	Surveys of Internet usage carried out over a 6-week period	6-month surveys might reveal neural damage, or loss of concentration and loss of ability to undertake 'deep reading'

9.10   toning down the strength of an affirmation: 1

Mark the following in terms of how strong / weak they sound.

(a) *Too strong.*

(b) *OK (i.e. good for the purposes of not appearing too convinced, too certain or arrogant).*

(c) *Too weak.*

1. It may be the case that these findings could possibly find an application in …

2. Other researchers may benefit from …

3. Other researchers should use these findings to …

4. Our findings prove that …

5. Our findings suggest that …

6. These findings will certainly be useful for …

7. These findings would seem to suggest that in certain circumstances there might be a possibility to …

8. This would seem to indicate that …

9. To the best of our knowledge this is the first time that …

10. We believe that these results show that …

11. We hope that other researchers will …

(1) c (2) b (3) a – in this context *should* almost sounds like an obligation (4) a (5) b (6) a (7) c (8) b (9) b (10) b (11) b

9.11   toning down the strength of an affirmation: 2

Insert appropriate words / phrases into the spaces. More than one word / phrase may be possible for the same space.

Words / phrases: although, as far as we know, generally, potentially, likely, possibly, probably, this could / might / may be, we believe that, would appear / seem

(1) _____ many authors have investigated how Ph.D. students write papers, (2) _____ this is the first attempt to systematically analyse all the written output (papers, reports, grant proposals, CVs etc.) of such students. The results (3) _____ to demonstrate that students from humanistic fields produce more written work than students from the pure sciences and this is (4) _____ due to the fact humanists are (5) _____ more verbose than pure scientists.

1. although

2. we believe that / as far as we know

3. would seem / would appear

4. probably / possibly / likely [US English], this could / might / may be

5. generally / potentially

9.12   toning down the strength of an affirmation: 3

Where possible insert Phrase A into the phrases below. Where this is not possible, insert Phrase B.

Phrase A: As far as we know … *or* To the best of our knowledge …

Phrase B: We believe that …

1. _____ there are no studies in the literature on this topic.

2. _____ principal agent theory has never been applied to …

3. _____ no attention has ever been paid to the risks of …

4. _____ our results show that X = 1.

5. _____ this is the first time that x has been shown to be equal to 1.

6. _____ the findings reported in this paper clearly indicate that …

(1) a (2) a (3) a (4) b (5) a (6) b

9.13   toning down the strength of an affirmation: 4

Underline the phrases in the Discussion below that readers (including editors and referees) might consider too direct or arrogant. Then paraphrase the third paragraph using your own words and adopting a more neutral stance.

Our analysis of English scientific papers written now and in 1960 shows conclusively that the language has become considerably simpler and more readable. Sentences have become shorter and more verbs and personal constructions are being used. This clearly means that both native and non-native readers find English increasingly easier to understand and digest.

The results contained in Tables 4 and 5, and in Figure 1, substantially bear out our initial two hypotheses. Firstly, that English is not inherently more simple than other European languages; secondly that the inherent complexity of many European languages is reflected in the difficulty of carrying out certain bureaucratic tasks in the countries where those languages are spoken.

Consequently those authors who state that English is in fact simpler must clearly have been basing their findings on erroneous data. In fact, many have made the unforgivable mistake of comparing 21st century English bestsellers with more erudite 20th century literary works from other European languages. They also erroneously equate simplicity with a lack of complexity. Our findings highlight with a 100% level of certainty that such a correlation is unfounded, and what is more, it has led millions of students around the world to mistakenly believe that English is in no way a difficult language to acquire.

Paragraph 1: **conclusively** should be removed from the first sentence, and **clearly** could be removed from the third sentence.

Paragraph 2: this does not contain anything too strong

Paragraph 3: this is extremely critical of other authors and is likely to offend both referees and readers. Below is a paraphrased version.

In our opinion, the reasons behind the conclusion by other authors that English is a simpler language are due to the fact that they compared literary works from different decades (English works from the first two decades of this century, and other European works from the last three decades of the last century). We would also like to call into question the correlation that is often made between simplicity and lack of complexity. Our findings would seem to show that a language can be both simple and yet have high levels of expressiveness. Unfortunately the simplicity / lack of complexity correlation has led many non-native learns to think that English is easy to learn.

9.14 direct versus hedged statements 1

Below are extracts from two different Discussions. Choose between a direct statement (the first word in bold) and a hedged statement (the second), as appropriate. In some cases both would be equally valid.

The experiments reported in our study indicate that there (1) *is / may be* an interplay between bilingualism and the ability to solve problems (Figure 7). Learning two languages simultaneously (2) *is / could be* the first signal to alert the brain that there are always at least two ways to approach a problem. In fact it is known that similar signals increasingly (3) *accumulate / could accumulate* in the brain and then (4) *trigger / may trigger* additional neuron connections. Subsequently the connections (5) *form / may form* a supra-linguistic system, which (6) *replaces / can replace* the normal monolingual neural network (Smith and Jones 2010; Startski and Utch, 2012).

However, the exact signal responsible for problem-solving in relation to language has (7) *not been identified / to the best of our knowledge not yet been identified.* We (8) *found / believe we have found* a cDNA encoding for a putative signal. We (9) *we have thus managed / it would thus seem that we have managed* to identify the missing key step in the neuro-linguistic process. Our findings (10) *will / may* find applications in many diverse areas, for example language teaching in schools, critical thinking and logic systems.

1. *may* – this is just the author's finding, she cannot be sure that her finding is correct.

2. *could* – this is just the author's finding.

3. *accumulate* – already established fact in the literature.

4. *trigger* – already established fact in the literature.

5. *form* – already established fact in the literature.

6. *replaces or can replace* – this is from the literature, however **can** may be appropriate if it means that it has the potential to replace but does not always necessarily do so.

7. *to the best of our knowledge not yet been identified* – this is the author's belief.

8. *believe we have found* – this is the author's belief.

9. *it would thus seem that we have managed* – this is the author's belief.

10. *may* – the author believes her findings have the potential to be applied. If she had used *will*, it would indicate that she is 100% certain about such applications, which would sound rather arrogant.

9.15   direct versus hedged statements 2

The following is an example from the Discussion section of a paper entitled The Archeology of Water in Gortyn, by archeologist Elisabetta Giorgi. Her research has revealed what she believes to be a new perspective on Roman aqueducts. She takes the specific case of Gortyn, the most important Roman town on Crete. Until now it was believed that the basic function of the aqueducts in the Roman period of history was to transport water into towns for use by individual citizens in their homes. However, Elisabetta hypothesizes that the main function may have been to provide water for fountains and thermal baths. There are no Romans around today who can confirm her hypothesis, so she cannot be 100% sure of the validity of her findings. Consequently, she needs to 'hedge' her claims.
Choose between a direct statement (the first word in bold) and a hedged statement (the second), as appropriate. In some cases both would be equally valid.

We calculated that the minimum amount of water supplied (1) *was / appears to have been* around 7,000 m^3 per day. On the basis of demographic estimates for that century, people (2) *consumed / may have consumed* from 25 to 50 l per day. Yet our calculations (3) *show / would seem to show* that, if thermal baths and fountains are not taken into account, approximately 280 liters per head could have been pumped into the town. This figure (4) *is / would seem to be* 30 l per day higher than the daily average consumption of a post-industrial European country such as Italy. The quantity of water that (5) *flowed / might have flowed* along the aqueduct (6) *was thus / thus appears to have been* much greater than was needed by the population living in Gortyn, which has been estimated as being around 25,000. Therefore the aqueduct (7) *was built / was probably built* not exclusively to provide drinking water for the citizens. Other authors (8) *contend / would appear to contend* that Roman citizens (9) *had / may have had* running water in their houses and they cite findings at Pompeii as evidence of this. However, our previous archeological research into aqueducts in other Roman towns (10) *certainly highlights / would seem to indicate* that the aqueducts were not necessarily built for the benefit of common citizens.

For a more detailed explanation, see Sect. 9.14 in *English for Writing Research Papers*.

(1) was (2) consumed (3) would seem to show (4) is (5) flowed (6) thus appears to have been (7) was probably built (8) contend (9) had (may have had also possible) (10) would seem to indicate

Notes:

1–2, 4, 5, 8 are calculations or facts not interpretations

3, 6, 7 are hypotheses based on the author's research

9 both forms are possible here and would depend on what the authors in the literature actually stated in their paper. If they stated that Romans had running water, then the correct answer is *had*; if they only hypothesized that Romans may have had running water then the answer is *may have had*.

10 both answers are possible, but the second is an example of a hedge (i.e. the author is protecting herself from possible criticism).

9.16 discussing possible applications and future work

Many Conclusions sections end by stating possible applications for a meth-odology, and the future work that the author intends to do himself / herself or to throw open to the community. Choose two or three cases from below. Write the final lines of the Conclusions.

Example for 1
We have found the recipe for an elixir of life for animals. With this elixir, animals could potentially live forever, thus it could immediately be applied to animals that have terminal illnesses. Future work will involve testing the elixir on humans, and investigating ways to make simulations of their possible lifespan.

	YOUR METHODOLOGY	APPLICATIONS	FUTURE WORK
1	The recipe for an elixir of life for animals	Animals with terminal illnesses	Human testing
2	A methodology for the transmutation of base metals into gold	Metal-recycling plants	Stone into gold
3	Architectural bridge-making techniques (arch, cantilever, and suspension) have properties that can be used on a metaphorical basis to bridge gaps in understanding between people in conflict	Diplomacy, teaching, conflict resolution, counseling, psychotherapy	Investigate the transferability of other areas of architecture to non-engineering building activities
4	A prototype kaleidoscope showing different 3D views of the same object	Engineering, design, architecture	Reducing costs of production, reducing size
5	A means to reduce human weight through breathing techniques and brain exercises rather than diet and physical activity	Partial cure for obesity	Improve techniques and exercises

Section 10: Writing each section of a paper

The exercises in this final section are based directly on the advice given in the companion volume *English for Writing Research Papers* (see table below). The exercises ask you to write a section of your paper either by answering certain questions or by following a suggested structure. To do each exercise in depth would take you several hours. So consider focusing on the questions or parts of the structure that you think would cause you the most difficulty, or which you think you need to practise the most. No key is provided in this section, so your teacher will need to correct your work.

TOPIC	ENGLISH FOR WRITING RESEARCH PAPERS
Title	Chapter 11
Abstract	Chapter 12
Introduction	Chapter 13
Review of the literature	Chapter 14
Methods	Chapter 15
Results	Chapter 16
Discussion	Chapter 17
Conclusions	Chapter 18
Acknowledgements	20.17

A. Wallwork, *English for Academic Research: Writing Exercises*,
DOI 10.1007/978-1-4614-4298-1_10, © Springer Science+Business Media New York 2013

10.1 abstracts

Write an Abstract related to your current research, alternatively invent some research. Choose one of the two possible structures below.

STRUCTURE 1

1. Give a basic introduction to your research area, which can be understood by researchers in any discipline. (1–2 sentences).

2. Provide more detailed background for researchers in your field. (1–2).

3. Clearly state your main result. (1 sentence).

4. Explain what your main result reveals and / or adds when compared to the current literature. (2–3 sentences).

5. Put your results into a more general context and explain the implications. (1–2 sentences).

STRUCTURE 2

1. Begin by saying what you did plus introduce one key result, i.e. begin with information that the reader does NOT already know. (1–2 sentences).

2. Introduce the background by connecting in some way to what you said in your introductory sentence / s. (1 sentence).

3. Use the background information (which the reader may or may not already know) to justify what you did, and outline your methodology (and materials where appropriate). (1–2 sentences).

4. Provide some more information regarding your results. (1–2 sentences).

5. Tell the reader the implications of your results. (1–2 sentences).

10.2   introductions

Write your own introduction following the structure below. You may decide to leave out some of the stages.

1. Define the topic, suggest why it is important and of interest and / or give some brief historical background. (1–3 sentences).

2. Outline the accepted state of the art plus the problem to be resolved (i.e. the gap). (2–4 sentences).

3. State your major objectives, i.e. how you intend to fill the gap. (1–2 sentences).

4. Introduce the background literature that you intend to refer to in order to give the rationale behind your research. Ensure you make reference to current insufficient knowledge of your topic. For example, you may think a particular study did not investigate some necessary aspect of the area, or how the authors failed to notice some problem with their results. (an appropriate number of sentences).

5. Make a clear statement of how what you paper represents an advance on current knowledge, and what your objective is. (2–4 sentences).

6. Announce / Preview the main results of your work. (1–4 sentences).

7. Give the structure of your paper. (3–4 very short sentences).

10.3   creating variety when outlining the structure of the paper

The Introduction of a paper typically ends with an outline of how the paper is organized. How does the author of the text below create variety in his description of the structure? Then do 10.4.

For some years the community has encouraged collaborative clinical trials. In this section we describe the first of two unreported results on such trials that we believe deserve publication. Then, in Section 2, we outline the broad perspectives that have shaped the direction of the literature on clinical trials. Section 3 answers the question: 'Under what circumstances have trials been carried out since the introduction of the new norms?'. Finally, we draw some conclusions in Section 4. We believe that this is the first time that such an approach has been applied to an analysis of clinical trials.

10.4   outlining the structure of the paper

Write a description of the structure of a paper based on the following information.

The rest of the paper is organized as follows.

Section 2 – theoretical hypotheses based on x and y.

Section 3 – methodology and data sources.

Section 4 – results.

Section 5 – discussion + limitations.

Section 6 – conclusions + possible extensions of the analysis.

The rest of the paper is organized as follows. The second section presents the theoretical hypotheses, based on the economics of media markets and communication studies. The third section describes the empirical methodology and the data sources, while the fourth presents the results. The last section draws some conclusions, and discusses the limitations and the possible extensions of the analysis.

10.5   survey of the literature

Write a survey of the literature following this structure:

- Introduction to aspect 1 (i.e. one specific area of research within the field).
- Support from the literature regarding Aspect 1.
- Mini summary explaining how your work represents an advance on what is already known.
- Introduction to Aspect 2, and so on.

10.6 methodology / experimental

Write your Methods section by answering some or all of the questions below. Your first subsection may be a general overview of the methods chosen, how they relate to the literature and why you chose them. Then in each subsequent subsection you:

(a) *Preview the part of the procedure / method you are talking about.*

(b) *Detail what was done and justify your choices.*

(c) *Point out any precautions taken.*

(d) *Discuss any limitations in your method or problems you encountered.*

(e) *Highlight the benefits of your methods (perhaps in comparison to other authors' approaches).*

1. What / Who did I study? What hypotheses was I testing?

2. Where did I carry out this study and what characteristics did this location have?

3. How did I design my experiment / sampling and what assumptions did I make?

4. What variable was I measuring and why?

5. How did I handle / house / treat my materials / subjects? What kind of care / precautions were taken?

6. What equipment did I use (plus modifications) and where did this equipment come from (vendor source)?

7. What protocol did I use for collecting my data?

8. How did I analyze the data? Statistical procedures? Mathematical equations? Software?

9. What probability did I use to decide significance?

10. What references to the literature could I give to save me having to describe something in detail?

11. What difficulties did I encounter?

10.7 results

Write your Results section following this structure:

1. Highlight those results (including those from controls) that answer your research question.

2. Outline secondary results.

3. Give supporting information.

4. Mention any results that contradict your hypothesis and explain why they are anomalous.

10.8 discussion: 1

Write your Discussion section by answering some or all of the questions below.

1. Do my data support what I set out to demonstrate at the beginning of the paper?

2. How do my findings compare with what others have found? How consistent are they?

3. What is my personal interpretation of my findings?

4. What other possible interpretations are there?

5. What are the limitations of my study? What other factors could have influenced my findings? Have I reported everything that could make my findings valid or invalid?

6. Do any of the interpretations reveal a possible flaw (i.e. defect, error) in my experiment?

7. Do my interpretations contribute some new understanding of the problem that I have investigated? In which case do they suggest a shortcoming in, or an advance on, the work of others?

8. What external validity do my findings have? How could my findings be generalized to other areas?

9. What possible implications or applications do my findings have?

10. What further research would be needed to explain the issues raised by my findings? Will I do this research myself or do I want to throw it open to the community?

10.9 discussion: 2

Write your Discussion section by following the structure below.

1. Statement of principal findings.

2. Strengths and weaknesses of the study.

3. Strengths and weaknesses in relation to other studies: important differences in results.

4. Meaning of the study: possible explanations and implications for clinicians and policymakers.

5. Unanswered questions and future research.

10.10   differentiating between the abstract and the conclusions: 1

Read the two texts below very quickly. Decide which is the Abstract and which the Conclusions.

TEXT 1 In this work, MOR-ON, a tool for the prediction of the behavior of students at university is developed. MOR-ON is based on a lumped and distributed parameters approach and is capable of describing both the social and in-lecture behavior of first-year students. On the basis of the boundary conditions applied to the model, it is possible to obtain the operating map of changes in behavior. Particular care is devoted to the analysis of alcohol consumption and of its influence on exam outcomes. The predictive capabilities of our tool are evaluated by simulating a reference case: first-year students from the 1950s. The most important parameters for the description of the behavior are detailed and a set of these parameters are found, in order to accurately simulate the complete operating map. Finally, numerical results are compared to measurements and a good agreement between experimental values and numerical predictions is found. The study highlights that alcohol and recreational drugs are responsible for the moronic behavior of many university students, particularly in Anglo countries and northern Europe.

TEXT 2 This paper focuses on the development of a tool, MOR-ON for the prediction of the behavior of students at university. Based on a lumped and distributed parameter approach, the model is capable of obtaining the operating map of changes in behavior, without restrictions in the operating mode.

The social behavior of students is detailed along with how they behave while carrying out their academic tasks, for example during lectures and tutorials. MOR-ON evaluates the consumption of alcohol and recreational drugs and compares it to the students' exam results. The predictive capabilities are then evaluated by simulating the reference case of exam results from the 1950s and the relevant consumption of stimulants.

Finally, numerical results are compared to measurements and a good agreement between experimental values and numerical predictions is found.

The only significant difference between the two texts is in the last sentence of Text 1 which is a summary of the conclusions of the work. Both texts could be used as the Abstract, but neither lends itself very well to the Conclusions. Ensure that when you write your Conclusions, that they are not a cut and paste or paraphrase of the Abstract. The Conclusions are not just a summary of the paper, they should highlight the key results, quickly deal with limitations and implications, and outline paths for future research.

10.11 differentiating between the abstract and the conclusions: 2

The texts below describe a study by a researcher investigating when would be the optimum time for a female student to terminate her love relationship with her male partner. Compare the Abstract and Conclusions below, by answering these questions.

1. What information is given in the Abstract that is also given in the Conclusions?

2. What information is exclusive to the Conclusions?

3. How does the author use the current research in China to summarize the methods used in her South American research?

ABSTRACT Three red flags were identified that indicate that the time to leave him has come. These red flags are: five burps per day, two channel-zapping sessions per day, and fives games on the Playstation with friends per week. Many women have doubts about the right moment for leaving their partner. Often women wait in hope for a change in their partner's habits. One hundred couples (above all South American) were analyzed, recording their daily life for six months. Women were provided with a form to mark the moments of annoyance recorded during the day. Burps, channel-zapping sessions and games on the Playstation with friends produced the highest index of annoyance. The probability of eliminating these habits was found to significantly low when the three red flags had been operative for more than three months. Thus, these numbers provide a good indication of when the time to leave him has come. With these red flags, women will no longer have to waste their time waiting for the right moment.

CONCLUSIONS The three red flags identified in our research – numbers of burps, zapping sessions, and Playstation sessions – should enable women to understand when they need to leave their partner. To counter any effects due to the nationality of the women involved (predominantly South American in our sample), we are currently doing tests in China. The results we have so far for China would seem to confirm our initial findings, but with an additional fourth flag: time spent studying for examinations. In addition, the timeframe for the flags to be operative in China is two months, rather than the three months reported in this paper. We also plan to replicate our tests on a wider range of women and a longer time scale, thus increasing the sample base from 100 to 1,000, and increasing the recording of daily life annoyances from six months to twelve months. Future research could be dedicated to doing analogous tests to enable men to see the signs of when they should leave their woman, and for employees to identify when they should leave their current employment.

1. The main findings (three red flags).

2. Current research in China, *timeframe for the flags, sample base, length of time researchers spent recording daily annoyances,* future work.

3. By stating the differences between the research in South America and the new research in China. Note: these differences are highlighted in italics in the key to question 2.

10.12 conclusions: 1

Write your Conclusions section by following the structure below.

1. Revisit briefly the most important findings pointing out how these advance your field from the present state of knowledge.

2. Make a final judgment on the importance and significance of those findings in terms of their implications and impact, along with possible applications to other areas.

3. Indicate the limitations of your study (though the Discussion may be a more appropriate place to do this).

4. Suggest improvements (perhaps in relation to the limitations).

5. Recommend lines for future work (either for the author, and / or the community).

10.13 conclusions: 2

Write a paragraph summarizing one or more of the following points. The last few sentences should:

- Outline a general conclusion.
- Suggest some implications.
- Indicate lines of 'future work'.

1. Your government's performance in the last one to five years.

2. Your academic performance in the last year.

3. Your relationship with friends.

4. Your use of the Internet.

5. Your reading habits.

10.14   acknowledgements: 1

Write your Acknowledgement section by including some or all of the following.

- Sources of funds.
- People who gave significant technical help (e.g. in the design of your experiment, in providing materials).
- People who gave ideas, suggestions, interpretations etc
- The anonymous reviewers.

10.15   acknowledgements: 2

Think of all your academic achievements in your life so far. Write 50–100 words of acknowledgements to all those people who have helped you.

Acknowledgements

I would like to thank all my Ph.D. students as well as the following people for allowing me to reproduce (modified) extracts from their work: Matteo Borzone, Carlo Ferrari, Roberto Filippi, Elisabetta Giorgi, Estrella Garcia Gonzalez, Caroline Mitchell, NASA Scientific and Technical Information Division, Chris Rozek, Anna Southern, and Alistair Wood.

The text for exercises Sects. 6.9, 6.10, and 7.20 were taken from *Principles and Methods of Teaching*, by James Welton, published in 1906 by University Tutorial Press. The exercise on writing definitions (Section 8) is an adaptation of an exercise from *Study Writing* (Cambridge University Press, 1987), an excellent book by Liz Hamp-Lyons and Ben Heasley. The exercise on structuring the results section is based on recommendations by Maeve O'Connor in her book *Writing Successfully in Science* (Harper Collins Academic, 1991).

About the Author

Since 1984 I have been editing and revising scientific papers, as well as teaching English as a foreign language. In 2000 I began specializing in training Ph.D. students from all over the world in how to write and present their research in English. I am the author of over 30 textbooks for Springer Science + Business Media, Cambridge University Press, Oxford University Press, the BBC, and many other publishers. I hold short intensive courses for Ph.D. students and researchers on how to write and present their research.

I would welcome comments on improving this book, particularly the keys. Please contact me at: adrian.wallwork@gmail.com.

Editing Service for non-native researchers / Mentorship for EAP and EFL teachers

My colleagues and I edit, revise and proofread manuscripts for publication in international journals. We specialize in papers written by researchers whose native language is French, Italian, Rumanian, Portuguese or Spanish. With 30 years of experience in editing papers, we also offer a mentorship program for teachers who would like to learn how to enter and work in this interesting and very remunerative field. Contact: adrian.wallwork@gmail.com.

A. Wallwork, *English for Academic Research: Writing Exercises*,
DOI 10.1007/978-1-4614-4298-1, © Springer Science+Business Media New York 2013

Index

A
Abstracts 5.24, 10.1, 10.12, 10.13
Acknowledgements 5.31,10.14, 10.15
active vs passive 7.16, 7.17
adverbs 5.5, 5.6, 7.13, 7.14
ambiguity 6
arrogance 9.10-9.13

B
brackets 1.4
brevity 5

C
can 5.11, 5.12
capitalization 1.7, 1.8
causes 4.14
certainty 9.2–9.4
commas 1.1, 1.2
comparing 8.9–8.14
concessions 4.16
conciseness 5
Conclusions 5.30, 10.10–10.13
confirming evidence 8.5
consequences 4.6, 4.15
contrasting views 8.10
contrasts 4.10, 4.16
cutting and pasting 7.1

D
definitions 8.1–8.3
Discussion 9, 10.8, 10.9

E
effects 4.15
evaluations 4.11, 4.20, 4.21
Experimental 5.29, 8.11

F
figures 4.19, 5.21
findings 8.6-8.8
future work 9.16

G
generalizations 8.4
gerund 6.3, 6.4

H
he 6.8-6.10
hedging 9.10–9.15
highlighting 8.7, 8.8
hyphens 1.5, 1.6

I
impersonal expressions 2.3, 2.4
impersonal style 7.16, 7.17
importance of findings 8.6
-ing form 6.3, 6.4
Introductions 5.25, 10.2

L
level of certainty 9.2–9.4
limitations 9.5–9.9
link words 4, 5.14–5.17
linking sentences and paragraphs 4.1
Literature review 5.27, 5.28, 8.9, 10.5
long paragraphs 3.3-3.6
long sentences 3.1, 3.2, 3.8–3.10, 5.22–5.24

M
masculine pronoun 6.8–6.10
Materials and Methods 5.29, 8.11
Methods 5.29, 8.11

O
objections and alternative views 9.1

P
paraphrasing 7
passive vs active 7.16
personal style 7.16
plagiarism 7
political correctness 6.8-6.10

A. Wallwork, *English for Academic Research: Writing Exercises,*
DOI 10.1007/978-1-4614-4298-1, © Springer Science+Business Media New York 2013

possible applications and future work 9.16
prepositions 5.5
processes 4.13
punctuation 1

Q
qualifications 4.16
qualifying 9.6
quoting statistics 7.2

R
redundancy 5
rejections 4.16
repetition of key word 4.12, 6.1, 6.2
reservations 4.16
Results 8.6-8.8, 10.7
Review of the Literature 5.27, 5.28,
        8.9, 10.5

S
semicolons 1.3

solutions 4.17
spelling 1.11
structure of the paper 5.26, 10.3, 10.4
summarizing 7.17–7.20
Survey of the Literature 5.27, 5.28,
        8.9, 10.5
synonyms 7.7–7.14
syntax 2

T
tables 4.19, 5.21
time sequences 4.18
titles 5.4, 5.13
toning down 9.10–9.13

V
verb + noun constructions 5.8–5.13
we vs passive 7.16

W
word order 2